UPROOTED

UPROOTED

A GARDENER REFLECTS ON BEGINNING AGAIN

PAGE DICKEY

TIMBER PRESS

Portland, Oregon

Frontispiece: Rose hips, *Rosa nutkana*

Page 6: Young woods at Church House

Published in 2020 by Timber Press, Inc.

The Haseltine Building

133 S.W. Second Avenue, Suite 450

Portland, Oregon 97204-3527

timberpress.com

Printed in China

MIX
Paper from
responsible sources
FSC® C014688
FSC
www.fsc.org

Text design by Vincent James

Jacket design by Faceout Studio

ISBN 978-1-60469-957-9

Catalog records for this book are available
from the Library of Congress and the British Library.

FOR DM

[I am] in the condition
of a transplanted tree
which is hesitating
whether to take root
or begin to wither.

— ANTON CHEKOV

CONTENTS

LEAVING

We walked away from thirty-four years of making, nurturing, loving a garden. How many thousands of perennials and bulbs did we leave behind; how many flowering shrubs—hundreds? How many trees did we plant young and watch grow and stretch, becoming flowering umbrellas, casting shade, and turning fiery in the fall? We left behind the drifts of flowers that only happen with the passage of years—snowdrops, winter aconites, spring phlox in the woodland, foxgloves, verbascums, Johnny jump-ups, poppies, larkspur in the graveled gardens, lungworts spreading their haze of blue in the back of the borders.

I thought I would always live at Duck Hill, that it would be my last house. It wasn't big, a simple farmhouse, old and quirky, with small, book-filled rooms littered with down-cushioned sofas and chairs in faded linens and chintz, forgiving, frayed Oriental rugs on the floor. There were enough bedrooms in its rabbit warren to house our children when they came home, and, in later days, their wives and husbands, and their children. The kitchen, the heart of the house, was spacious, with a great old Garland stove, shelves of crockery, a long table and chairs in the middle, and doors opening onto the terrace and garden.

It is so American to turn our backs on a home and move on. And American too, that in a matter of months, it is all wiped away, those drifts of flowers, those fragrant shrubs, those young trees. Even the old farmhouse is torn down, obliterated, to make way for the structure that is the new owner's house. And now this strange place, no longer my home, is on the market again.

Why did we abandon Duck Hill? Our cost of living, for one reason. Our oldest son looked at our finances and said we needed to save money. And residing in New York's Westchester County, sixty miles from Manhattan, was expensive. Then the garden, if we were being honest, was becoming too much for us. Three acres of intensive plantings. Hedges to trim, borders that needed constant editing, roses to prune, flowers to deadhead, a large vegetable garden to tend, a woodland garden to care for, a meadow to weed! My husband, Bosco, was turning eighty and I,

Duck Hill in autumn

six years younger, was nonetheless finding the maintenance a challenge. Our garden had gained some notoriety over the years, and we welcomed a constant stream of visitors who wanted to see it during the growing season. The pressure of keeping the garden up to snuff, especially as the sultry heat of summer set in and it became junglelike, weighed on me.

We didn't want to hand our garden over to hired gardeners nor could we afford that as a solution. And our futile attempts at simplifying it were laughable. I replaced some of the neediest perennials in our mixed borders with shrubs. We talked about grassing over those borders and just leaving the hedges and boxwoods. But then there were all the gardens in gravel with wonderful seedings of flowers and bulbs—how could we possibly simplify those? In the end, we just gave up.

I felt stricken for a few weeks after we made the decision to move, periodically dissolving into tears as the enormity of leaving this world I created hit me. But as I slowly accepted the necessity of the move, I felt stirrings of excitement at the thought of finding a new nest. I began dreaming of the possibilities of a move—less garden, more land, another quirky old farmhouse, maybe an apple orchard, a real meadow.

We invited young avid gardening friends to come dig. Not knowing what time of year we would be moving or to where, we wanted pieces of our best plants to find homes elsewhere, insuring their survival. Double trilliums, yellow snowdrops, epimediums, my collection of anemones and anemonellas, odd-colored violets (apricot, palest yellow, clear pink), species daylilies, iris, roots of my treasured old roses—bits of these were dug up and parceled out.

The house went on the market and, at first, it looked like it might be sold to people who loved the garden. In the end, a neighbor bought Duck Hill for convenience, and we knew the garden was probably doomed. In the fall, before the contract was signed, we dug up a few divisions of perennials we felt sentimental about as well as some precious bulbs, and these were stored for the winter in pots at a friend's place, lined along their stone wall and covered with leaves. Adam Wheeler of Broken

Arrow Nursery came and took cuttings of our roses. But all the woodies, the shrubs and trees we had planted and loved, remained.

Our focus turned to dismantling the house, tackling thirty-four years of accumulation, rooms full of *things*. Many of which, I might add, we still own. We are not minimalists. I remember the day we moved, in the middle of December, looking back at the house, swept clean but desolate, stripped of pictures and books, bibelots, furniture, and rugs, and thinking, Okay, I can leave now. We are taking our home with us.

The main garden at Duck Hill

FINDING

A

PLACE

MOVING
NORTH

Our friends were bemused when we said we were moving north, out of New York and into New England. Surely most people our age head south? And as I look out my window on a January morning to freezing rain and sleet, I wonder, were we crazy? Perhaps. But Bosco and I love the change of seasons. I, for one, am more than ready to take a break from gardening in winter, relieved to be forced by frozen soil and single-digit temperatures to call it quits. What a pleasure it is to read books without the weeds calling, to bake cakes and slow-cook stews, to daydream about flowers. I have time to scheme about gardens, to change my mind multiple times on how to improve their design, to plot what plants to add next spring, what seeds to order. Yes, here we have dreary days of freezing rain and the inevitable danger of lurking ice. Yaktrax on our boots are a winter necessity. But we still get excited, childlike, by snow in winter, willing to cope with its inconvenience in exchange for its silence and shadows, loving those sparkling sapphire and diamond days that Bosco says remind him of the Alps.

The long winter months have everything to do with the thrill we experience of emerging spring, the first sight of skunk cabbage, tightly furled and speckled rust-red, the sweet smell of earth thawing. To witness the bravest of flowerings—witch hazels, hellebores, snowdrops, aconites, and the tommie crocuses (*Crocus tommasinianus*). Suddenly I cannot bear to be inside, overcome with excitement to see what is stirring as each day new bulbs and perennials emerge, some of them friends I forgot we had. As soon as the soil is dry enough, I have every excuse to be outdoors—cleaning up, cutting down, planting trees and shrubs, and sowing lettuces, sweet peas, and spinach.

Aquilegia canadensis

Sweet peas

I was astonished the first time I heard from my daughter-in-law about "June gloom" and "May gray" in Southern California where she lives, a weather pattern there in late spring and early summer of overcast days, fog, and drizzle. I thought, But those are my favorite months. I treasure our gorgeously verdant May and June, pristine in their freshness, before bugs, disease, and heat take their toll. Every gardener I know who lives in the Northeast wishes these two months were longer, not only because of their beauty, but also because this is our precious window of time when we lift, divide, and replant perennials, as well as put in new plants and sow seeds, before the humid summer settles in. Inevitably, before we have finished all our spring planting, a rainless week of 80- to 90-degree weather will hit, and we know our window is gone.

Moving north means cooler days and nights throughout the growing months (a chance finally to successfully grow sweet peas), although, with global warming, who knows how much warmer it is going to be. For now, with shade from our trees and a frequent breeze, Bosco and I are reveling in the long opulent hours of summer. Our dinners get later as the days get longer, and we linger in the garden at dusk, weeding or drinking a glass of chilled wine on the terrace, talking about our day, as the sun disappears behind the pine trees.

And how can we give up the brilliance of our Northeast autumns? Our landscapes, our distant hills and low wetlands, the woodland floor with its bounty of mushrooms, the maples that line our streets—all stun us with the hues of their autumn dress. Some years the show is better than others, always much discussed—whether it was the cold at night or how dry or wet the weather has been. But always, throughout October and into November, we are treated to an ever-changing panoply of rich colors that at times takes our breath away. Moving north to New England and a setting one zone colder promises an even more intensive show.

LOVING
NEW ENGLAND

Most of my childhood was spent in the suburbs of Philadelphia. Until I was twelve, we lived in the home that had belonged to my grandparents, a friendly three-story, stucco-clad house close to its neighbors, with a screened porch that ran the width of the house off the living room and, better still, a sleeping porch off an upstairs bedroom. My room, when I was no longer a baby, was up on the third floor, the leafy limbs of an old tree just outside my windows. Wide interior sills supported houseplants I somehow acquired, I don't remember how, and, at least for one spring, a fishbowl full of tadpoles I collected. My sister and I spent most of our time with a housekeeper named Mrs. S who had an ample lap to climb into and the patience to play Parcheesi with us and sit by at night while we listened to *The Lone Ranger* on the radio and colored in our coloring books.

There was an old garden in back of the house, enclosed within a high privet hedge, where a pattern of beds cut into lawn was filled with tulips and, later, summer phlox, the peppery fragrance of which still sends me back to childhood. A shiny-trunked 'Kwanzan' flowering cherry tree at one end of the garden offered shade over a gathering of cushioned wooden chairs and chaises. Oddly, I have no memory of lingering in that garden, though obviously I had buried my nose in the phlox, but I see in old photographs my parents sitting under the tree, having cocktails with friends in their summer whites. I do have a vivid recollection, sometime in the late 1940s, of planting white and yellow corn with my father and sister in the small vegetable garden on one side of that privet hedge, and the pleasure of thrusting those fat seeds into the tilled earth in long strict rows.

By the time I was a teenager, we had moved farther out in the suburbs to a somber fieldstone house in a neighborhood of clashing azaleas and pink and white dogwoods. Our loving Mrs. S was replaced by a strict and humorless Norwegian widow named Mrs. Wig, who idolized my father but thought two teenage girls were a nuisance at best. My sister and I unkindly called her "Wig-Wag" behind her back, willingly ate her hot baking powder biscuits with crabapple jelly for breakfast, then walked the mile to school on a road without sidewalks, in spring pulling the inchworms out of our hair.

My mother was never easy with children, and I have little memory of interaction with her growing up. She took care of the household accounts, ordered groceries on the phone (she once said she wished she could take a pill instead of eating), and in her spare time worked on double-crostics from the *Saturday Review*, played Solitaire, and read murder mysteries, which she called bloods. She loved her whiskey, the result, she used to say, of being a Prohibition baby. She had been a beauty, and, as a young couple, my mother and father were quite glamorous, frequenting the opera and black-tie parties, dazzling onlookers on the dance floor. But as she got older she began to fall a lot—suffering from a loss of balance exacerbated by drink—and she became increasingly reclusive. My father was a dreamer, passionate about his books and about art and music, a passion he willingly shared with his two daughters whom he adored. But he never found his calling in life, quit work in his forties to write a book on Florentine Renaissance art that was never published, and, for a time, buried his disappointment with evenings spent away from home, late morning risings, and, from noon on, the companionable clinking of ice and pouring of whiskey highballs with my mother. Unfortunately, there was no driving need for him to work, for they lived comfortably enough on money my mother inherited. I remember an underlying melancholy at home when I was a teenager, which might explain my dislike of the house and the place.

Goldenrod

And yet they were loyal parents, always ready to support my sister and me in our times of crisis.

My sweetest memories of growing up are the times spent with my mother's older sister, my beloved Aunt Helen, at her small eighteenth-century Cape Cod home in Hingham, Massachusetts. She was lovely looking as a young woman and accomplished, graduating from Wellesley College with honors in 1923, where she developed a love of classics, history, and poetry. She never married but had many friends, and was active all her life supporting the arts and land conservation. In many ways she was the opposite of my mother, industrious but reserved at a large party, with no interest in the trappings of society. Her beauty had faded somewhat when I knew her, her figure short and generously wide, a Beatrix Potter of a woman, who dressed sensibly in woolens for the country life she led. Mornings with Aunt Helen were spent walking with her two bearded collies on trails she carefully maintained through 30 acres of mature oak and beech woods behind her house, and it was on these walks that I learned from her the names of wildflowers and birds. I remember one warm day when I was quite young, resting with her on a fragrant bed of pine needles in the woods while she taught me the Lord's Prayer. I am agnostic at best, but I have to this day a fondness for that prayer.

Mornings at Aunt Helen's were also for baking bread, a sensuous occupation I learned to love on those visits, and meals were inevitably delicious, thought out and celebrated daily, using vegetables and fruit from her garden. For she was a gardener, and I mostly remember spring borders full of lungworts, Jacob's ladder, and Virginia bluebells, interspersed with myriad small bulbs that had naturalized—scillas, chionodoxas, grape hyacinths. I remember, too, sitting with her on the terrace outside the kitchen shelling peas, popping in my mouth as many as I dropped into the colander. I first dabbled in drawing and painting with watercolors in Hingham with my aunt, an occupation she enjoyed throughout her life. Evenings were spent reading out loud after dinner,

E. S. Nesbit when I was young, *The Five Children and It* and *The Phoenix and the Carpet*. Later there was Austen and Trollope to fill our nights.

One day in her mid-eighties, Aunt Helen called me with the triumphant news that she had just climbed the old maple that bowed its limbs over ledge rock by her driveway. She said she'd been thinking of doing it for years. On her last visit to Duck Hill in her early nineties, we walked slowly around the garden, her arm intertwined with mine. "Oh, Pagie," she said, "if I were just a little younger, I'd have you dig me up a piece of this and that."

I bring this up because my aunt was my mentor; she taught me about everything I still love in life, from dogs to books to food to gardens to paths in woods. But also because I think it was at her home in Hingham, spending time with her there, that I first knew that I loved New England.

Precious summers, however, were also spent with my sister at our family's summer place in New Hampshire on a rock-bottom, spring-fed lake called Squam at the foot of the White Mountains, where loons serenaded us at nightfall. I designed and planted my first garden there before I discovered boys, creating a path to a circular clearing in the woods where the septic tank was apparently buried. I laboriously edged that path and that circle with granite stones I gathered from the property, and then planted the circle with wildflowers I dug up from the surrounding forest. I shudder to think of the trilliums and lady's slippers I might have displaced in my ignorance.

For several summers as a young married woman with four small children, my then-husband and I rented an old farmhouse in Sandwich, New Hampshire, that sat in a field of wildflowers. There was no television in the house, we read aloud at night, and the rooms were filled with bouquets I cut of the black-eyed Susans, goldenrod, and meadowsweet outside all around us. We packed picnic lunches most days to have on wide, warm granite ledges at a local rapid river that had pools of deep water into which the children leapt for hours. It was affectionately called the Pothole, and was little known until *Life* magazine named it one of the ten best swimming holes in America. At night, at least once a week, we would all go square dancing in the next village of Tamworth. Those were treasured days.

Bosco visited New Hampshire as a young teenager shortly after arriving in America as a refugee from Hungary and remembers going square dancing there too, much to his delight, having only been versed in foxtrots and the waltz. And he spent a vacation waiting tables at an inn a few miles from our lake retreat the year I was twelve and he was eighteen, which began his life-long addiction to swimming. (As far as we know, we didn't meet, although my family frequented that inn, and it's possible he waited on us but would hardly recall a young sprout of a girl.) He went on to college in the Berkshires, and that cemented his fondness for New England.

THE NORTHWEST CORNER OF CONNECTICUT

Duck Hill, where I spent the prime of my life, and where Bosco joined me for the last fourteen of those years, was in North Salem, New York, teasingly near the southern edge of New England. But the village had characteristics similar to those over the border—clapboard farmhouses lining its streets, a valley threaded by a rock-strewn river, high patchwork fields edged in tumbled stone walls, apple orchards. In the seventeenth century, much of North Salem was, in fact, part of Connecticut until that state surrendered it to New York in a swap for more valuable land elsewhere. New England was our nostalgic destination when we decided to move from Duck Hill, but not too far away, we thought, and someplace where we already had friends.

The place we chose was the northwestern corner of Litchfield County in Connecticut, nestled beneath the Berkshire Mountains, a few miles from the Massachusetts border and just to the east of the Hudson Valley of New York. A little over an hour away, this setting had been in the back of my mind for several years as somewhere I could bear moving to if I ever had to leave Duck Hill. I was enchanted by the area's pastoral beauty, its openness, the constant views of hills—sometimes distant, other times startlingly close, the rolling terrain and sudden rock outcroppings, the steep pastures of grazing cows. The Housatonic River, flowing south from its source high up in the Berkshire Mountains, cuts through this picturesque valley, wide and quick-running, past farms and small villages clustered with eighteenth- and nineteenth-century clapboard houses and churches.

Queen Anne's lace

Northwest Connecticut countryside

There was a time then when I would not have been enamored of this landscape, because the hills there were gradually being denuded. A booming iron ore business in the region was sustained by the plentiful local quarries of limestone and richly forested hillsides. A huge quantity of charcoal was needed in the blast furnaces to smelt iron ore, and consequently all the surrounding mature forest was cut down. How sad, how barren it must have looked. At the time, there were visions of Northwest Connecticut becoming a great industrial center. But those hopes were dashed as the Erie Canal and the railroads opened up the west and iron was made more efficiently there. The great stone forges were abandoned and the Berkshire foothills gradually became reforested with what is now a mature second growth. With the boom over, the villages never expanded, remaining bucolic if uneconomical, as industrialization passed them by.

I began coming up to this corner of Connecticut four decades ago because of gardens, first visiting Linc and Timmy Foster's remarkable woodland garden in Falls Village. The Fosters were famous for their writings about rock gardening; Lincoln's invaluable book, *Rock Gardening*, was published in 1968 and reprinted in 1982 and contained a wealth of information about wildflowers as well as alpines. My old copy is worn with use. Their 6-acre garden at Millstream House was a destination for sophisticated alpine enthusiasts as well as young burgeoning gardeners like myself. I remember being entranced by the corydalis in yellow and white (*Corydalis lutea* and *C. ochroleuca*) frothing and tumbling over rocks that bordered their fast-running stream, and primroses everywhere, and sheets of forget-me-nots among the ferns and wildflowers new to me.

In the 1990s, I came up once a year to visit Fred McGourty and his wife Mary Ann at their Hillside Gardens in the town of Norfolk, which Fred proudly called "the icebox of Connecticut." It all started with Fred's book, *The Perennial Gardener*, in which, among humorous anecdotes, he shared his vast knowledge of perennials. I loved the book and wrote to tell him so. We began to correspond and soon started exchanging visits. "Come visit this summer, but let me know ahead so we can have iced tea on the terrace and debate the merits of cardoons and Scotch thistle and other good things in life," Fred wrote in spring 1989. And so I went, returning home with a car full of new-fangled perennials from the irresistible nursery that lined his driveway.

Connecticut's Northwest corner—Salisbury, Sharon, Falls Village, Norfolk, Cornwall—remains rich in gardens and gardening, bringing us back and back to this idyllic place at the foothills of the Berkshires, acquiring friends each time we visit. How fitting it seemed for us to live here. Now, with Duck Hill on the market, we began our hunt.

If you are a nest-builder as I am, searching for a home is a wrenching, emotional experience. Especially if you love where you are, the home you're leaving behind. We were fortunate to have a sympathetic realtor, Laurie Dunham, to guide us, a friend foremost, honest, caring, and full of laughter. Countless prospective houses in the five designated towns were discarded just by looking at their details on the internet—too pretentious, too costly, in deep dark woods, smack on a busy street, too small, too big, not old enough. We knew we wanted an old house, one with character and history, though Bosco leaned more to Victorians with high ceilings and those impossibly steep staircases, remembering with fondness the gatehouse in Irvington, New York, that he gave up when he married me, as well as apartments in Paris where he lived for seventeen years and brought up his children, and his childhood home in Hungary where vast rooms were warmed by tall, ceramic-tiled stoves. I, on the other hand, yearned for the coziness of farmhouses with seven-foot ceilings and intimate rooms. We wanted interesting and varied land, of course, the more the better as far as I was concerned. But Bosco wisely insisted that we settle somewhere not too far away from neighbors and a town with a grocery store as well as other amenities. I knew our new home would involve some compromise.

We became enamored of a house in Sharon early on, before Duck Hill was sold. It wasn't great looking on the outside, having a bay window at one end, which I loathed, with land that was either deeply shaded by a grove of Norway spruce or suspiciously wet: A large stand of phragmites—that dreaded invasive wetland species—was not a good sign. To make matters worse, a kidney-shaped pool (the shape, my bête noire) had been placed directly outside the back of the house; you almost fell into it as you opened the door to the yard. But the inside utterly charmed us. Most of the house was indeed old, with wide-planked pine floors and handsome moldings, charming fireplace hearths, low-ceilinged, cozy rooms beautifully enclosed and divided by bookshelves. A small but well-equipped kitchen led to a master bedroom and bath on the ground floor and a tiny, winding back staircase took you to three charmingly quirky upstairs bedrooms. How the grandchildren would adore those stairs, I thought, rather adoring them myself. A spacious screened porch jutted out from the house, and, most important for Bosco, so did a recent addition of a high-ceilinged, elegant living room that you stepped down into, filled with light from that bay window.

We started to conjure up living there, what we could do to make the land more tenable, plotting where our furniture would go, even in our heads hanging pictures on the walls (one of Bosco's favorite occupations). I looked up on the internet what to do to get rid of phragmites. We went back with our realtor to walk the thin wooded edge of the property, hacking our way through a ground cover of barberry in search of some decent native trees. Our plan was to bid on the house after Duck Hill was sold—we did not want to own two houses. But, as so often happens, in the meanwhile someone else bought it. Thank goodness, for I realized in retrospect that some of its charm was due to its decoration by its talented designer-owner, the kitchen was too small, the situation and the shape of the pool unbearable, and the land not at all what we dreamed of having. I hate to think of the mosquitoes.

We went back three times to a substantial house with spacious rooms in Salisbury before we admitted to each other, sitting on a bed in a guest room upstairs, that we hated it. Again, its land was without merit (Norway maples the only trees and acres of brambles), and there was something bland about the place: it didn't sing to us. It's amazing how, sometimes, you try to talk yourself into a house, because you desperately want to latch onto a home. The beauty and the trouble with Northwest Connecticut is that it is still quite rural and there are very few houses, let alone ones on the market. We walked away from that Salisbury house dejected and worried. Duck Hill was now under contract.

CHURCH
HOUSE

NOT A
FARMHOUSE

It was late in the day, we were tired, but we were in the neighborhood, and so we decided with our realtor to take a look at a place I had previously dismissed after seeing its specs on the internet. Although reasonably priced, the interiors appeared without charm and the house seemed to sit right on a road. None of us had actually been to the place and we got royally lost trying to find it, even though it was just minutes from Main Street in Falls Village. What we saw as we finally came up the driveway and stepped out of the car was a starkly simple house set in what seemed to us the most beautiful landscape.

The land was sunlit, open, a wide lawn in front transitioning into a meadow that gently dipped down to pine woodland. A great and glorious sugar maple towered over the house, grounding it, giving it gravitas. Other venerable trees graced the lawn, among them an ancient, picturesque apple tree that spread its boughs at the end of the drive-way, and the largest black cherry I had ever seen, rough-barked and lichen-covered with limbs that writhed as if out of an Arthur Rackham drawing. All this was splendid, but what was most unexpected, what took our breath away, was the view of distant hills. The house looked out onto Canaan Mountain, part of the southern Berkshires, which rose dramatically above the woods in soft swoops and was painted that late October afternoon in lavender and burnt sienna. We were done for.

The house was not what either of us had in mind, not a Victorian or a farmhouse. It was a meeting house, built in 1793 on this "sightly spot" as the Battle Hill Methodist Church, the first Methodist church to be erected in New England. "I have circulated a subscription for the building of a church here," wrote Freeborn Garrettson in 1790, who with his friend "Black Harry" Hosier, was welcomed to Falls Village, or

Canaan as it was known, on a summer day and preached to about 500 people in the open air. "Harry preached after me with much applause," Garrettson added. Remarkably, Harry was a freed slave who was a popular traveling preacher due to his musical voice. He could not read but knew his passages of scripture by heart.

The meeting house those Methodists built is almost Shaker in aspect, a simple, high box, steeply roofed, but without a steeple, clothed in clapboard, originally with stout, fluted Doric-style pilasters at its corners. From an old photograph taken, I would guess, in the early nineteenth century, an entrance door appears on the south façade, and a small tree has been planted to one side. I wonder if it is the great sugar maple. The inside in the early 1800s was described as "having plastered walls, good seats, and a stove." At this time, a gallery was apparently added upstairs. The present driveway by the east front of the house (what I took for a road in the realtor's pictures), was, indeed, at one time a main through-road over Battle Hill, called the Canaan-Salisbury Highway, informally known as the Old Stagecoach Road. Luckily for us, it was discontinued sometime in the late 1800s.

Preachers practiced here until the early 1850s when people began to complain about climbing the hill and how difficult the old building was to heat. By then, the center of business in town was on Main Street, close to the new railroad and the great Housatonic River that ran below it. That's where you wanted to be. And so, a new Methodist church of white-painted clapboard was built on Main Street in 1855, this time with a fine square steeple and a series of tall graceful windows.

The old church on the hill stood empty for many years. A newspaper article written in 1908 speaks of a plan to restore the edifice to its original appearance, "even to the most minute detail," for it was one of the town's oldest historic buildings, and "stands as a monument to the faith and zeal of the earlier settlers." As far as I can make out, that restoration never happened. In 1930, the church was bought by the Martin Murray and Jesse Fife families, the grandparents and parents of

Marie Hewins, who was born and grew up here and lives in town today. She says her grandmother and her great uncle did all the work to make the church into a two-family home, her grandparents living upstairs, her parents downstairs. The house remained in Marie's family until her parents died in 1988, when it was sold to a real estate broker who had plans to develop the property. The cost of putting in roads turned out to be prohibitive, the plan was thankfully abandoned, and the house sold to a bachelor who enjoyed it as a weekend retreat.

The entrance was changed to the east front near the driveway when it was made into a house, and either the broker or the bachelor added large, mullioned picture windows on the ground floor and a series of small, paned windows on the second floor, which give this façade of the house a rather cross-eyed look and jarred dreadfully with its historic architecture. But, oh, what wonderful views those windows offer.

Dazzled by the land and those views, we went inside the house that autumn afternoon to see what it offered. I knew from the internet that the living room was a narrow galleylike space, too cramped to have a comfortable cluster of seating. A respectable front hall divided the living room from the kitchen, which was also long and narrow but had a good stove and a small eating area with two of those picture windows looking out to the fields and Canaan Mountain. Upstairs, curiously, the ceilings were considerably higher (the former gallery?) with a sunny landing, floor-to-ceiling bookcases, and that series of windows facing east that afforded a breathtaking view. Three ample, sunny bedrooms and two large baths were reached off that landing. The attic had two rooms that could be studies or a hangout for grandchildren, with some extra room for stuff.

My heart sank a bit as I walked around and saw that few if any vestiges of oldness had been preserved indoors. No quirky details, no beautiful old moldings remained. A fine wooden bannister led the way up the stairs, satiny with use, dating at least from Marie's grandparents' time. Hand-cut beams in the attic and the cellar spoke of the building's

The view at Church House

age, but the original floorboards had at some point been replaced, and only two windows (upstairs) seemed old, perhaps from church days, with wavy panes of glass. At Duck Hill we were lucky enough to have three working fireplaces with charming old mantels. Here, there were none.

We went home puzzling how to make the house more livable, drawing sketches on paper, and wondering how an addition could be achieved without disturbing the integrity of the old church. Our friend Bunny Williams, who masterfully designs interiors for a living and lives just around the corner in Falls Village, stopped by, saw what we loved about the setting, and suggested a plan for the house: Add a living room with a fireplace at the end of the hall on axis with the front door, make the existing long, narrow living room into your bedroom, and build a glassed-in mudroom off the kitchen. And that's what we did.

RENOVATIONS

Duck Hill sold in December 2014, and, early in 2015, we bought the old meetinghouse on Battle Hill and dubbed it Church House. We knew with the impending construction that we would probably not be able to move in for a year, so, except for essential clothing, a few books, and a couple of favorite pans for cooking, all our belongings went into storage. Bunny and her husband John Rosselli, with their historic generosity, offered us their guesthouse as a place to live—this knowing that we had two dogs in tow, our old terrier, Roux, and our lurcher, Posy, about the size of a small horse.

We asked Peter Talbot, the architect from Washington, Connecticut, who designed the Boscotel, Bosco's one-room Greek Revival house that was his retreat at Duck Hill, to collaborate with us on the new addition. We settled on a high-ceilinged living room very much in the same style as the Boscotel, with coffered panels overhead, a fireplace at the far end, and tall windows that wrapped around the room on the south, west, and north sides, flooding it with light. Directly above each window, Peter added a narrow clerestory window, recalling the eyebrow windows often seen on old houses here. At night, when we are reading or watching TV in this room, I often catch glimpses of the moon through one of these west-facing clerestories.

Bosco always complained at Duck Hill that there was nowhere to hang pictures, since almost every vertical surface was covered with bookshelves and consequently books, much to my liking. Now he dug in his heels and, backed by our architect, said there would be no bookshelves in the new living room. I meowed about how much warmth and interest books add to a room, to no effect. The narrow walls between the windows were all to hold paintings and drawings, period.

Between the original front hall and the new living room, there are two small rooms—or hallways, for the living room is not attached

The living room at Church House

to the old church, but floats away from it—linked by these narrow, low-ceilinged spaces, similar to the passageways that lead from an old New England house to a barn. Within this enfilade of rooms, I have my library, the walls packed with books, a ladder ready to reach the top shelves, a comfortable chair and lamp for reading in one corner, small Oriental rugs on the floor. Favorite novels, biographies, and literary letters reside in one open room, with books on history, nature, animals, art, and design in the other. A small section holds well-loved children's books from my childhood. Only gardening books are missing: Except for the piles by our sofa in the living room, all others are relegated to the generous floor-to-ceiling shelves on the second floor landing where another upholstered chair and a table are placed, ready for research and perusing.

I remember when we finally moved in, just before Christmas in 2015, after all the kitchen paraphernalia and crockery was unpacked and put away, the clothes hung and the beds made, and the furniture arranged, I faced the long and pleasurable, if exhausting, task of emptying box after box of books. As I became reacquainted with each volume, turning it over in my hands, dislodging any lingering dust with a blow, then setting it in its rightful place on a shelf, I listened to the background music of tap, tap, tap, as, with hammer and nails, Bosco hung his pictures. When, finally, the walls were all decorated with drawings and paintings and the shelves richly filled with books, the house settled in as our home.

At Duck Hill, the usual entrance to our old farmhouse was through a glassed-in, brick-floored porch facing southeast where we had plants growing year-round. In winter, old iron plant stands were full of cyclamen, and two stout tables held Bosco's collection of begonias along with clivias, amaryllis, and South African bulbs (mostly lachenalias and veltheimias) that we forced into bloom. As winter stretched on, the tables and floor became littered with pots of daffodils, as well as crocuses, scilla, and other small bulbs pulled from our cold frames.

The library

We planned the same sort of room at Church House as our back entrance, a place for boots and gardening shoes and walking sticks, as well as plants and muddy dogs, in this case connected to the kitchen on one side and the garage on the other, where we thought we might actually house a car. Halfway through the construction of the mudroom, I thought how nice it would be to have a long narrow dining table in the middle of the room, offering another surface for displaying plants, but, more to the point, an alternative place to eat meals when we were more than six. For even eight at our round kitchen table was snug, everyone elbow to elbow. I measured and measured, knowing we had barely enough space, as the width of this long room had been dictated by our desire to save a beautiful unidentified viburnum just outside, perhaps a cultivar of *Viburnum trilobum*, laden with clusters of translucent red fruit in the fall and winter.

As it turned out, the viburnum was destroyed one day by heavy machinery used to work on the foundation (why we didn't dig it up ahead of time and heel it in somewhere safe, I do not know) and our mudroom could have been a whole lot roomier. Merely one of the inevitable mistakes we made along the way. We do have meals out there for a gang, surrounded by potted plants, and, in summer, Bosco and I eat breakfast and lunch at one end of the table with the doors, facing east and west, wide open so we can watch hummingbirds and goldfinches that come to feast on our catmint and sunflowers in the gardens just outside.

Landscape with white-flowering crabapples

A LANDSCAPE INHERITED

Every day of that year living in Bunny and John's guesthouse, we coaxed Posy into the car (our terrier, Roux, was old and usually stayed behind) and came over to the house. In the winter our visits were brief, a chance to walk some of the 17 acres we now owned and let Posy run. As spring came and the construction started, we often spent much of our day here. After the obligatory walk through the house to greet the carpenters and admire the progress, we spent the rest of our time getting acquainted with our land.

From that first autumn day when we discovered the place and fell in love with the setting, I realized that the landscaping around the sides and front of the house was unusually sophisticated. Tall spruce trees, planted in a staggered line on either side of the beginning of our driveway, completely shielded us from our neighbors and the busy road. In front of this dark green screen, a lavish sweep of white-flowering crabapples was established on one side of the drive, and groves of multistemmed shadblows and Japanese lilacs on the other. A low, stone wall defined a narrow garden along the front of the house, and, here, two more shads in tree form cast light shade. Shads, or serviceberries as they are sometimes called, varieties of *Amelanchier*, are one of my favorite native understory trees, graceful in habit and lovely in flower and fruit. But I rarely see them used in quantity around a house.

On the north side of the lawn and meadow, a drift of six kousa dogwoods fronts the woods, decorative even in winter with their mottled gray and tan trunks. A small summerhouse built in the 1950s, which is no architectural thing of beauty but serves as our screened porch in summer, stands close to the garage on this north side, and is shielded from view by large bushes of fragrant *Viburnum carlesii*. A heavily fruited crabapple, a variety, I learned, called 'Ralph Shay', masks it from the driveway, and two gnarled old apple trees frame it on the east and west sides. Beyond this structure, out of sight from the house, someone built a rectangular swimming pool (happy Bosco!), edged in granite slabs set in lawn and surrounded on two sides by banks of early and late-flowering hydrangeas. How clever, I thought, to place a pool where you weren't even aware of its existence for the six months of the year when it is covered and unused. Who did all this fine placement and planting, I wondered that autumn day.

We soon learned that our bachelor owner hired the garden designer Nancy McCabe about twenty years ago to create the plantings we see around the house and also to designate and design a pool. This was a delightful coincidence, because I knew Nancy and admired her work, so much so that she and her designs are featured in a chapter of the book *Breaking Ground* that I wrote twenty-three years ago, maybe just before she was asked to work at Church House.

That first year we mostly watched and schemed how to enhance and improve what Nancy had left us once the construction was done. Meanwhile, we put bluebird boxes in the east field and placed two cushioned rattan chairs on the lawn facing the high grass and the woods. We sat there often in the afternoons while work on the house continued, enjoying the bird life that by spring was abundant. If the days were hot and buggy, we sat in our screened summerhouse, which we more often called the pool house, reading books and answering emails. I started exploring our fields and woods, sensing that this wild land would become an absorbing passion.

In the back of the house, facing west, where much of the construction was taking place, lawn quickly relaxes into meadow, then spreads out to a frame of more woods. At the far end, just before the woodland edge, there is a curious small limestone quarry, so hidden from view until you are upon it that it reminds me of a ha-ha. To our horror, twenty-one full-grown ailanthus trees had taken hold on one side of the quarry. In the Northeast, ailanthus, along with the Norway maple, are our worst exotic invasive trees. One of the first tasks we took on, in the late summer before moving in, was cutting down those trees, and as they were felled, brushing their stumps with Roundup. I knew this was probably necessary since ailanthus stalks rise with a fury when you cut the trees down, but we were off to a rocky start in my quest to have an organic garden. We are still pulling seedlings of ailanthus from the field, but less and less every year.

There are some beautiful slabs of rock and marble in the quarry, and immediately I thought what a wonderful hidden garden it could be, a rock garden of all sorts of dianthus perhaps, or of native scramblers and creepers—sweetfern; *Rosa virginiana*; St. John's wort, *Hypericum frondosum*; and New Jersey tea. A friend suggested we scatter seed of the delicate, white-flowering *Penstemon digitalis* among the ledges. *Campanula rotundifolia* already flowers in cracks of the rocks, and clumps of ginger-colored little bluestem are colonizing one section. But *Rosa multiflora* abounds there, and bittersweet clothes its high banks in menacing thickets. These invasives have to be dealt with before any extensive plantings can succeed. After we moved in, I put my dream of a secret garden in the quarry on the back burner, while I concentrated on more visible flower beds and borders around the house.

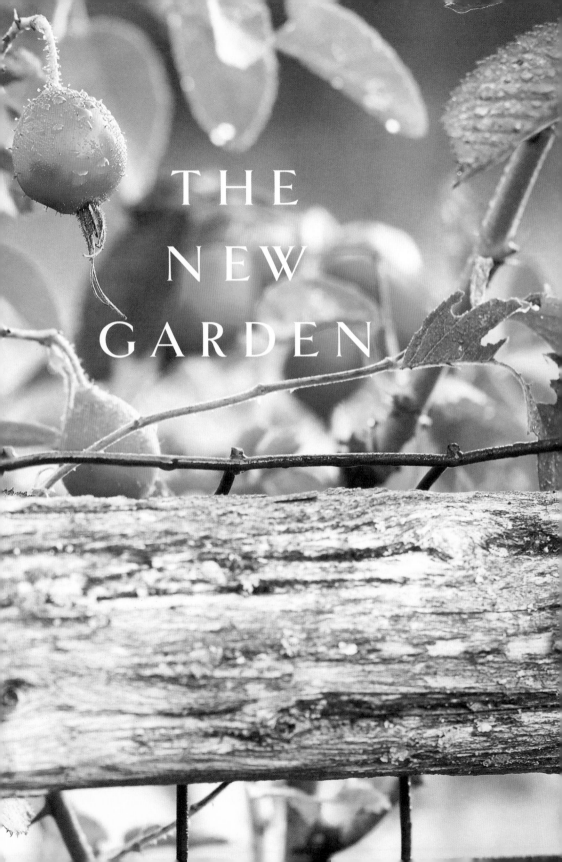

THE
NEW
GARDEN

Hedges and boxwood at Duck Hill

THE FRONT BORDERS

Shortly after we settled at Church House, I posted a picture on Instagram of the view from the house—that wide open expanse of sky above green lawn relaxing into meadow, a statuesque tree in the foreground, a rim of pines in the distance, the soft Berkshire hills rising in the background. One friendly follower, who obviously knew the garden at Duck Hill, commented on the post: "I can't wait to see what you're going to do with this." My reaction was, No, No, No! I'm not going to do anything with it. No hedges, no carving up the lawn into garden rooms, no cluttering what seemed to me a perfect scene.

We love how our new house sits in its setting, surrounded by fields, the openness all around, the sense of space and sky, altogether different from Duck Hill, where the necklace of hedged garden rooms around the house confined it, gave it a closed sort of intimacy. A wise friend, visiting Duck Hill many years ago, warned me not to give up all the negative space on our 3-acre parcel. I ignored him and did just that, filling every inch of ground with trees and shrubs and hedges, boxwoods and flowers. It was all pretty and I loved it, but it offered no rest to the eye. We are lucky now to have a generous amount of land, and I hope the small gardens clustered around Church House do not intrude on the bigger, wild beauty that surrounds us and that I treasure.

I have discarded the idea of hedges in those small gardens, as much as I love their geometry and the sense of surprise they afford, feeling somehow they wouldn't be appropriate here. I want the borders and beds around the house to have a certain transparency, the gathering of flowers to include meadow denizens that dance and wave, through which we catch glimpses of the fields, allowing a connection with the wild land. This transparency I crave goes against rules of garden design

Ironweed, sunflowers, and Russian sage in the front garden at Church House

I have for years advocated—that you don't put flowers in front of a view, that you don't want to see everything at once, that you do want to create a sense of mystery. But the mystery at Church House happens as you walk *away* from the gardens, down the sun-filled meadow paths into the dark of the woods, and there are astonished to come upon bluffs and a ravine, vernal pools, lichened rock outcroppings, sheets of native sedges, and delicate wildflowers, unexpected and sometimes unknown, at least to me.

We inherited two gardens from Nancy McCabe, one surrounding the pool, the other against the east front of the house, five steps down from the front door, and three broad steps below the mudroom. Nancy built a 2-foot-high dry stone wall along the length of the house and 20 feet out from it, then dug a long flower bed within that space cut through with a stone path. On the east side of the wall, between it and the driveway, she found room for another narrow border.

Peonies, which blessedly survive decades untended, still dot the whole length of the beds on either side of the wall, becoming its most common thread, in various shades of pink, perhaps a little too much pink. On the driveway side of the wall, *Geranium dalmaticum* runs along the ground, not a flashy cranesbill but attractive in leaf whatever the weather and smothering what would otherwise by now be a host of weeds. Clumps of iris—pale blue and purple bearded sorts and slender Siberian cultivars—have lasted years of neglect, good verticals now in need of division and replanting. A large baptisia flowers splendidly by the driveway in June, and a grove of silvery Russian sage, *Perovskia atriplicifolia,* is showy in summer and fall. I could never get Russian sage to stand up at Duck Hill, but here it billows nicely, and if it leans slightly into the driveway, that's okay. Because you first glimpse this narrow border along its length as you drive or walk up the driveway, I thought it cried out for a repetition of plants to be most effective.

I added three more clumps of Russian sage, valuing its gray stalks and lavender flowers that carry on for many weeks, impervious to drought or disease. I repeated the native, shrublike *Baptisia australis* along the border, loving its luminous pea flowers and glaucous foliage, its easy appropriateness and all-season good looks. Stretches of low-growing *Amsonia* 'Blue Ice', with inky purple buds and long-blooming sky blue flowers, now dress the front edge at intervals, as well as *Nepeta racemosa* 'Walker's Low' and some clumps of shrubby lavender, which loves our limey soil. Even in the dead of winter, the lavender is telling, its foliage a gun-metal gray in contrast to the buff and brown of asters added to the border.

Tall *Boltonia asteroides* var. *latisquama* 'Snowbank' now leans over the peonies with clouds of asterlike white daisies in autumn, and the October-blooming aster native *Symphyotrichum oblongifolius* 'Raydon's Favorite' is a haze of blue on either side of the front steps. I've dug in white and purple alliums among the perennials here, the tall pom-pom sorts, and will try to add to them yearly to guarantee a good display. Like tulips, even a few alliums add a good punch to a border. Unlike tulips, they are deer proof, a concern here since we have no deer fence and, at this point, no deerhound or lurcher either. Our Posy died of old age shortly after we moved in.

I'm not sure I would have designed flower borders smack up against the front entrance and kitchen entrance to the house, virtually a cottage garden at the two doorsteps. But here it is and I find something charming about it—it lightens the seriousness of the old church. The beds are manageable in size, the porous, sweet soil already dug and enriched, though with an underlying and sometimes impenetrable base of limestone. It's a place where I can indulge my desire to grow some perennials, intermingled with a few shrubs and bulbs, in a moderately sized space.

Nancy laid down rectangular slabs of old granite as the stepping stones through the garden on the house side of the wall, and, although they were handsome, I found them awkward and uneven for walking. They are now replaced with pea gravel paths edged in natural stone, my favorite medium for garden paths. The driveway we inherited was a mess of pitted and chipped asphalt, made more ragged by the construction. Because it ran right along the front garden, dividing it from the lawn and field, and was therefore an important element of the scene, we covered it with the same gravel, at first re-asphalting the surface, then tarring it, and laying down the pea stone so it would (mostly) stick.

Nancy's garden between the wall and the house was virtually destroyed by the construction, although peonies remained against the wall, and I managed to save some white veronicastrum to return to the beds. For woody structure, I planted another standard shadblow to cut the horizontal line of the new mudroom, and added some dwarf Korean lilac and a cluster of *Clethra alnifolia* 'Ruby Spice' near the kitchen for fragrance. A tea viburnum, *Viburnum setigerum*, is now in the corner by the kitchen, a sort I love because of its grace and drooping red fruit. But it looks a bit spindly here, being quite narrow and vase-shaped in habit, and I know it looks best when planted in a grouping of four or five. Two tall lilacs remain at the corners of the house, and the front steps are bordered with large, old, somewhat ratty boxwood bushes that I am hoping to revive with dressings of compost and occasional light clipping. I have added a few more box bushes for winter green and bones, but the rest of the plants here in the front garden are herbaceous, and many are natives.

One of my goals as I add flowers to this garden is to encourage birdlife. Already the berried shadblows attract flocks of cedar waxwings in late June, and they return in fall for the viburnum fruit. The baptisias that I've added to all the front garden beds are favorites with hummingbirds, and *Nepeta sibirica* 'Souvenir d'André Chaudron', which I rescued from the construction, now throws up 3-foot flower spikes by the kitchen steps in June, attracting sometimes hourly visits from a hummingbird. Penstemons are a draw also, and, later in summer, *Salvia guaranitica* 'Black and Blue' brings the hummingbirds back. I've allowed *Helianthus* 'Lemon Queen' on the edge of this garden, a beloved thug that grew in abundance at Duck Hill, introduced here because the goldfinches love it.

Rudbeckia subtomentosa 'Henry Eilers' takes up a large space in mid- to late summer, a favorite of mine with its billowy branches and clear yellow, quilled rays of daisy flowers, and it, too, is a favorite with the finches. The only trouble with this rudbeckia is it tends to sprawl, especially if we have fierce rainstorms, which we seem to a lot. And yet it looks stiff and awkward when staked and trussed. I am going to give it the Chelsea chop next May, shearing it back by half, and see if that helps. The only problem is it might then not bloom until October.

I am partial to tall plants and ones that sway in the wind, so I've added more of our native Culver's root, *Veronicastrum virginicum*, in white and pale blue, which satisfies on both accounts. It is an elegant vertical with whirled leaves and spires of tiny tubular flowers that bees feast on for both nectar and pollen. Tall ironweed, *Vernonia baldwinii*, is here too, with vibrant purple heads of flowers in August that bring giant swallowtail butterflies fluttering and dancing in pairs and threes above the blooms. Where the beds are sunniest, I've added clumps of *Amsonia hubrichtii*, the thread-leaved blue star that is especially prized for its golden yellow leaves in fall. But the taller *A. tabernaemontana* is here too, with its showier cymes of light blue flowers in late spring and upright stalks of leaves so fresh and pleasing that I often cut them for fillers in bouquets. I've repeated yarrow, the bright yellow *Achillea* 'Moonshine', near the front of the border, its disc-shaped blooms a fine contrast to most other flower shapes.

Farther along the beds where there is half-shade, downy skullcap, *Scutellaria incana*, threads its way among kalimeris, daylilies, and the pale yellow, tiny-thimbled foxglove, *Digitalis lutea*. I was first introduced to our native skullcap in a planting Dutch garden designer Piet Oudolf created in the shadowy bosque at the Battery Park in lower Manhattan. He uses it in vast sweeps under the high-pruned sycamores there, and I instantly coveted it for its pretty, loose racemes of two-lipped, lavender, white-throated flowers. A member of the mint family native to the eastern United States, it is happy in sun or half-shade and blooms for a long

The front garden in July

time in midsummer. Piet also uses a lot of bowman's root, *Porteranthus trifoliatus*, in that garden, and I now have clumps of it here in one of the shadier spots. I've always loved this native's delicate, spidery white flowers flaring out from red calyces.

On either side of the gravel path that runs through the front garden, I've again repeated catmint, *Nepeta racemosa* 'Walker's Low', along with the lady's mantle that was already here in abundance, and cranesbills, mostly the blue *Geranium* 'Rozanne'. The dwarf New England aster 'Purple Dome' brightens the edges in the autumn, and calamintha adds a milky haze of blossoms. Creeping Jenny, *Lysimachia nummularia* 'Aurea', lingers from Nancy's days, indeed creeping into the path, and its chartreuse foliage is welcome here. A few clumps of yellow variegated hakonechloa have been added to the shadiest edge of the garden, repeating that chartreuse.

Tulipa 'Tarda', that beguiling wild tulip that opens its starry yellow petals flat-faced against its leaves, now spreads its cheer along the edges of the gravel path in early spring. All the species tulips are especially valuable in a garden where maintenance is a consideration, for they tend to be perennial unlike the tall Dutch hybrids we know to require replanting year after year. So far, the chipmunks and voles have left the wild tulips alone. I've dug in some bulbs of the spring colchicum, *Bulbocodium vernum*, which blooms surprisingly early, its strappy magenta petals emerging from the bare earth before many perennials have flushed out. Snowdrops that we brought from Duck Hill now flower in March and April on either side of the stone wall. *Allium christophii* opens its starry softballs among edging plants in June, and the 3-foot-tall drumstick allium, *Allium sphaerocephalon*, mingles with larger perennials midborder. This curious member of the onion family sports a small, maroon and green egg-shaped head, and when used generously is wonderfully effective, like upside-down exclamation points, among early summer flowers and grasses.

I've started adding daffodils to the beds, delicate, pale favorites like the elegant, small cup 'Segovia', and snow-white 'Thalia', a triandrus narcissus with three exquisitely flared and pendant white flowers to a stem. A few clumps of the creamy jonquil 'Sun Disc', tucked around catmint, carry daffodil bloom late into the spring. These daffodils, along with miniatures such as 'Toto' and the dandelion lookalike 'Rip van Winkle', are best seen in close quarters rather than en masse in a lawn or field. I've made notes in my garden journal to plant many more narcissus in this garden, small drifts in the very back of the beds where, after they finish blooming, their yellowing foliage will be masked by burgeoning perennials. All these spring bulbs have a wonderful way of enriching mixed borders while taking up astonishingly little room.

It's going to take a few more years to tweak this small front garden to my liking. My notes are full of directions to move this and plant that. And perhaps it will never knock anybody's socks off. But it gives us pleasure, an intimate space of flowery instances through which we walk many times a day and, even when indoors, see just outside our mudroom door and windows.

Japanese anemones

THE
POOL ENCLOSURE

The second flower garden we inherited from Nancy surrounds the swimming pool. It is enclosed with post and rail fencing made out of rustic cedar trunks and branches (I suspect from our property), then faced with chicken wire. The fence joins the old summerhouse on the south side, making it perfect as a pool house, and cedar gates open into the garden from the south, east, and west sides. Hydrangeas are the theme that Nancy established here. Our native smooth hydrangea, *Hydrangea arborescens*, with corymbs of cream and white lacecap flowers, runs along the west length of the fence, interspersed occasionally with its far showier cultivar, 'Annabelle'. About midway, they are interrupted by a large clump of the taller, late-blooming peegee hydrangea, 'Limelight', sporting great fat plumes of white flowers in August and September. On the north side of the fence, 'Limelight' reappears backed by the elegant narrow flower panicles of *H. paniculata* 'Tardiva', together successfully masking the pool equipment.

I am heartily pleased to inherit these tough, well-loved shrubs, mainstays of the summer and autumn garden here in the north, offering an amazingly long succession of blooms. The cymes of *Hydrangea arborescens* transition from green to white to beige, the panicles of the later peegees open white, then deepen to pink before fading finally to a papery buff-tan. They are a lavish background for the pool and provide months of bouquets for the house.

The east length of the pool fence backed a wide border that was a weed patch when we arrived on the scene. Weeds, and, in one section near the pool house, in the half shade of an old apple tree, a carpet of pink-flowering Japanese anemones, probably the hybrid called 'Robustissima', so named for good reason. In an effort to lessen the anemone's hold on this part of the border where I hoped to add more hydrangeas, I dug and moved bits to other parts of the yard (young divisions transplanted most successfully), and gave flats of it away, only to have the rhizomatous plants knit right back together again. Dare I call this plant a thug? It is impossible to dislike: the basal rosettes of dark-green, grapelike leaves are handsome all summer, and the September-blooming, 5-petaled flowers, although a musty, mauvy pink, are held gracefully on their tall stems and are elegant in the garden or in a vase.

I found out from Nancy that this wide east border was originally a long narrow vegetable garden for her client, safely fenced in from deer. But now, I thought, adding hydrangeas to this space would give the garden continuity. For the strongest and simplest impact, I would have just repeated what varieties were already here. But the collector's gene is strong, and it is too tempting and fun to include some other cultivars of hardy hydrangeas with different colorings and flower shapes. I did add more 'Annabelle' and 'Limelight', but also the curious, low-growing *Hydrangea arborescens* 'Green Dragon' with lacy, fern-leaved foliage. The deep pink *H. paniculata* cultivars 'Quick Fire' and 'Fire Light' are settling in at the back of the bed, and, mid-border, the compact and early- blooming 'Little Lamb' and 'Dharuma' are fighting it out with the anemones. All these plantings are still small and insignificant, but I hope as they grow up they will complete the hydrangea frame in this garden.

Because the east bed is wide and empty earth still shows among the new hydrangeas, the temptation is irresistible to add odd perennials brought home from plant sales as well as slips of roses from friends. The soil here is deep, the sun is full, the garden is protected from deer—how could I not? The apothecary rose, *Rosa gallica officinalis*, already commandeers a swath, a gift two years ago from Deb Munson, a dear gardening friend and neighbor. Other rooted divisions of old shrub roses are tucked in mid-border, too young yet to start running when they will spell trouble. Summer phlox, *Phlox paniculata*, has a home here, the mildew-free white-flowered 'David', which I rescued

The pool garden

from the front garden during construction. I have recently added the small-flowered, magenta-pink 'Jeana', named the best-performing garden phlox at the Mt. Cuba trials and said to attract more butterflies than any other variety.

Every time I see a new species or cultivar of sanguisorba we don't own, I lose all control and want to bring it home. Thirty years ago, the only burnet I knew and grew was our native *Sanguisorba canadensis*, which thrived in a damp, half-shaded section of my garden. In late summer, wiry stems supported curious, off-white bottlebrush blooms 4 to 5 feet in the air above clumps of untarnished scalloped leaves. Now, thanks to plantsmen like Ed Bowen in Rhode Island, all sorts of burnets are easily obtained, among them the Asian species *S. tenuifolia*, *S. obtusa*, and *S. hakusanensis*, the latter two with fuzzy, arching pink catkins. You can have a long season of burnet blooms, for varieties of these hardy perennials are in flower from July through October, with decorative basal clumps of pinnate foliage and thimbles or catkins colored deep burgundy, pink, or white. This is a see-through plant, one that you can have in the front of a bed or mid-border, because the delicacy of its tall stems and bobbing flowers only enhances more robust flowers around them. I have three or four sorts in the cutting garden, for they are a delight in mixed bouquets, and several more crammed into the east bed of the pool garden.

It will be a good thing when the hydrangeas bulk up and visually start to knit together, because, at this point, this east border is just a mish-mash of plants. I have abandoned, at least for now, the attribute I know is most visually satisfying in a garden—simplicity. Last spring, however, I dug in plugs of *Geranium macrorrhizum* 'Album' along the edges of both the east and west beds in an attempt at more coherence. If it thrives, this cranesbill will make a fine ribbon of fragrant scalloped leaves and pink-tinged white flowers all spring and summer.

One of the first tasks Bosco and I took on, even before we moved into Church House, was to site and make ready a place for composting.

We decided on the outside of the pool fence facing north, next to the pool equipment and hydrangeas. There was enough room for three sections, about 6 feet square each, that we could contain with posts and wire. The idea was to have one of those sections in use, collecting leaves, garden refuse, and green scraps from the kitchen until it was full, then rest that pile while we filled a second section; and, by the time *that* section was full and resting, we were dumping on the third, and, in the first section, we were returning the compost, our black gold, into the garden beds. We do not tend to buy fertilizer for the garden, relying on compost to enrich the soil.

The view of the treasured compost heaps is unsightly from the pool area, although I have a picturesque idea of pumpkins and gourds growing there (note to try that). To mask the piles, we planted three *Calycanthus ×raulstonii* 'Hartlage Wine', a hybrid of our native *C. floridus* and the Asian *C. chinensis* that we had at Duck Hill and admired for its spidery, magnolialike burgundy flowers and bold, paddle-shaped leaves that enlarge and become almost tropical-looking by summertime. Sweetshrub, Carolina allspice, and strawberry bush are all fond common names of this fragrant shrub, especially in its native form. The flowers smell of "green tea and Damson plum preserves and occasionally of strawberries," according to Helen Van Pelt Wilson in her book, *The Fragrant Year*, though more "a fruit salad perfume of strawberry, banana, mango, and peach" to William Cullina, as described in his book, *Native Trees, Shrubs, and Vines*. Suffice it to say, the fragrance is deliciously fruity. Calycanthus gets to be a good 8 feet high, and will balance the tall peegee hydrangeas next to them but be a more effective screen. Although the main flowering is in spring, sporadic flowering continues through summer and I have cut a blossom or two in the fall to add to a posy.

Mixed east bed in pool garden

The living room terrace

THE CUTTING GARDEN

When the building of the new living room was finished, we found we had to grapple with a drop in grade on the south and west sides of the new structure. We settled on a dry-built retaining wall made of local stone, similar to the one in the front of the house but 2 feet taller. We sited the wall 20 feet out from the south-facing foundation to make room for a graveled terrace that could be entered from a small porch off the living room.

Clusters of cushioned chairs and tables now litter the terrace in summer, and large pots of annuals—blue and purple violas in spring, followed by lavender alyssum, blue or white scaevola, and white petunialike million bells (*Calibrachoa*) in summer. Bosco's figs are here, and sky blue butterfly bush (*Rotheca myricoides* 'Ugandense'), brought out from the garage where they are kept over winter. Smaller pots of white and purple pineapple lilies (varieties of the South African native *Eucomis*), spidery Peruvian daffodils (*Hymenocallis*), white acidanthera (*Gladiolus murielae*), and geraniums are staggered on the porch steps. The fragrant *Daphne ×burkwoodii* 'Carol Mackie' now flowers in a bed along the south wall of the living room, flourishing here in the hot sun, its cream-edged green leaves handsome all season.

A dear gardening colleague, Katie Ridder, arrived at Church House the first spring after we were settled with a bucket of rooted divisions of a rose I gave her when we knew we were moving from Duck Hill. It was the lovely, low-growing single white burnet, *Rosa pimpinellifolia*, given to me many years ago by beloved garden friend Robin Zitter, and I was thrilled to have it back. Once you have this burnet rose, you have plenty to give away, for, on its own roots, it runs with enthusiasm. I decided to plant Katie's rooted bits of burnet rose in the narrow west bed of the terrace, where it would be confined by the living room wall at its back and somewhat restricted by the stone and gravel in front. Odd perhaps to have a rose as foundation planting, but I knew it would be attractive in all seasons. Its tiny, fernlike leaves look fresh from spring until frost, and the flowering in May is lavish and sweet-smelling and is followed by dark plum hips in late summer. In winter, the bristly canes of the rose are colored a rust-red.

I thought at first we needed a tree off the southwest corner of the new terrace to shade it from the summer sun, and, with dreams of a magnificent magnolia or stewartia in my head, went out with a young helper to test dig a spot. And another, and another. Nothing doing. The ground diagonally south and west of the terrace wall in a radius of 50 feet is solid ledge rock. I marvel at the great maple that towers over the old church on this south side—wondering how it found the soil to survive for the past 150 years. Maybe, after all, that ancient tree doesn't need any competition. The sugar maple shades the new terrace all morning; and we find in the heat of summer we don't linger here at high noon, preferring to gather at the day's end, relaxing with our glass of wine, as the sun dips down behind the western pines.

I knew when we bought Church House I wanted a garden where I could grow some lettuces and annuals for cutting, nothing as large as the vegetable garden we had at Duck Hill—we were going to buy tomatoes now, not grow them—but similarly fenced in with a pattern of beds and paths. The disturbed area just below the new terrace on its west-facing side seemed ideal, with full sun from late morning until sunset. As the wall was being built, we had our good mason fashion steps down to the ground level, and, with string and stakes, I marked out a garden about 30 feet square. Too small, said Bosco, and in retrospect he was right, but I didn't want to add too much to my gardening workload.

An old birdbath from the herb garden at Duck Hill seemed right as the centerpiece of two crossing paths in the middle of the garden. Secondary paths created middle beds, 8 feet long and 4 feet wide, small enough to weed without stepping into them. I again used gravel for the

The cutting garden

paths, hoping that annuals would eventually seed in them. We decided on a rustic fence around three sides of the garden (the fourth side bound by our wall), using cedar from our woods, which we faced with wire to keep the critters out. Little did we know that the four-foot wall we had just built would seemingly overnight become a vole condominium, with breakfast and dinner at their beck and call. Tulips beheaded, seedlings devoured, violas demolished. I wish I had a video of the white-blooming tulip as it was tugged by its stem into the wall while I watched in disbelief. I had to laugh, before I cried. We bought a load of snap traps, recommended by our friend, garden guru Margaret Roach. And we put them in small boxes, each with one entrance hole, which we placed strategically along the foot of the wall. That first season, a lot of voles met their end in those boxes.

I hate killing these round-eared field voles for they are adorable-looking, and, unlike the white-footed mice that live in our leaf-littered woods, meadow voles don't carry Lyme disease. But protecting my plants brings out the beast in me. Although any bug found inside the house is carefully carried and released outside unharmed, I think nothing of squashing a Japanese beetle between my thumb and forefinger if it's found disfiguring one of my roses. (I leave one or two cans of soapy water out of sight under the rose bushes to knock the beetles into, which works well in the morning or at dusk when they are sleepy; this is my preferred, and certainly more genteel, method of doing in this enemy.) With the vole problem in the cutting garden, a more humane solution would be to seal the bottom 2 feet of the wall with cement; for, unlike chipmunks, voles do not seem to jump or climb any height. We found if we kept our big pots of violas on the terrace 6 or 8 inches away from the wall, they were untouched. We also have a superb rodent catcher on board—our adored mutt Sadie. She has an uncanny ability to find voles belowground and proudly brings us daily offerings.

Sadie, our vole catcher

In the outer beds of this garden, I've planted a few perennials that are good for cutting but also complement the annuals growing here. *Sanguisorba tenuifolia* 'Henk Gerritson' is one of them, a favorite burnet with 2-inch thimbles of dark burgundy that nod and dance on tall, wiry stems. It grows in front of tithonia, the Mexican sunflower, that I plant every spring along the back of the border against the stone wall. By late summer, the tithonia has topped the wall at 5 feet tall and is covered with large, vibrant, orange daisies that are favorite landing places for monarch butterflies. The flowers cut well and I use them in kitchen posies with red-striped yellow dahlias and zinnias like *Zinnea elegans* 'Queen Lime Orange' and 'Zinderella Peach'.

Am I in the minority loving orange? I suspect so. A touch of orange can jazz up a discreet sweep of lavender-blue and white, and downright dazzles with purple, its opposite on the color wheel. How wonderfully the brilliant orange, jewel-like butterfly weed, *Asclepias tuberosa*, enlivens a field of wheat-colored grasses and Queen Anne's lace in the wild. Orange flowers in the cutting beds keep company with all the other hot colors I love to cut for the house—fire-engine red, deepest wine-brown, and every shade of yellow from gold to palest lemon. Cooler hues are in the minority here, but enough in evidence to leaven all the hotness—the blue of borage, nigella, and forget-me-nots, the white of cosmos and feverfew, pale pinks, too, and purple. Two bushy stands of perennial anise hyssop, *Agastache* 'Blue Fortune', throw up spikes of dusty lavender-blue for many weeks in summer, complementing the tithonia and an edging of tender *Agastache* 'Apricot Sunrise' in one bed, and mingling with tall scarlet zinnias and dahlias in another. A patch of hardy geum edges the bed nearby, a hybrid variety called 'Sangria', with single to semidouble flowers, pure scarlet with a center of yellow stamens, looking like little roses or the blooms of strawberries. The flowers seem to float on wiry, 2-foot stems above basal clumps of prettily scalloped leaves. I've always loved this spring perennial, whether in orange, yellow, or red, although it didn't flourish at Duck

Hill. I kept losing it and stubbornly planting it again, virtually treating it like an expensive annual. 'Sangria' appears to be particularly vigorous at Church House and so far has survived our erratic Zone 5a winters, when the temperature in a day can go from –18 to 50°F.

Two statuesque perennials hold down the corners on the western side of the fence. One is a repeat of my favorite rampant sunflower, *Helianthus* 'Lemon Queen', and the other is *H. giganteus* 'Sheila's Sunshine'. Both these sunflowers have an abundance of delicate yellow flowers on gracefully swaying stems. 'Sheila's Sunshine' is remarkable, however, because she waits to flower until October and, if you haven't pruned back the stems in early summer, rises to 8 feet, like a tall, narrow shrub. The small, creamy blooms are alive with masses of bees right up until frost. I always try to have a few annual sunflowers in the garden for their drama in bouquets, especially the rusty brown and dusty pink sorts, and the pale yellow, smaller flowered 'Italian White'. Cosmos is a must in the garden, an annual I prize for cutting, especially the tallest sorts like white 'Purity'. They are enchanting gathered in a tall pitcher or vase, mixed with sunflowers and hydrangeas or just by themselves, and outdoors they add a winsome delicacy to the beds, their wide-cupped blooms moving with the least wind. Dahlias are added to the beds after all threat of frost is over and offer great punches of color and pattern, as do zinnias. But they are both stiff in habit, and benefit from the company of dancing colonies of cosmos.

The cutting garden is also a trial ground for flowers I've never grown before, in large part thanks to the young plantsman Helen O'Donnell, who owns and runs Bunker Farm in Vermont, which specializes in unusual annuals. Helen's expertise, her catholic choice of plants, is influenced in part by her month-long forays in England in past years, working as a student first at Hidcote and then at Great Dixter. Every year, we have a celebrated plant sale in our new neighborhood called Trade Secrets, and Helen comes down with flats and flats of sophisticated annuals, many of which are strange to me. She willingly describes

August in the cutting garden

Tagetes 'Cinnabar'

the unknown plant, and I come home with an embarrassment of new annuals to try. Because of Helen, I now always include *Emilia coccinea*, the tiny scarlet tasselflower. Its beguiling bright red tufts on slender 1-foot stems weave among stouter annuals and perennials at the front of the beds. The tasselflower is also a treasure in small bouquets, mixed with diminutive striped zinnias and nasturtiums. Helen introduced me to *Centratherum intermedium* 'Pineapple Sangria', the Brazilian button— now a favorite front-edge annual with purple, slightly shaggy, buttonlike flowers above serrated leaves that smell of pineapple. It is in flower from early summer to frost. The marigold cultivar, *Tagetes* 'Cinnabar', with Great Dixter in its provenance, is now coveted mid-border for its rich, dark, burnt-orange pinwheels with yellow centers.

The four narrow center beds of the cutting garden are used for seed sowing in the spring. This is where I start cosmos, sunflowers, amaranth, and tall zinnias that will mostly be transplanted to the outer borders. At least one bed is reserved for the rows of lettuces, mizuna, and arugula that I sow as soon as the soil is dry enough to handle in spring, usually sometime in early April in my area. I repeat sowings through summer as often as possible and hope to have a row growing in fall until frost. A few rows of leeks keep company with flowers in another bed. I grow ample amounts of Italian parsley, for I love to use it generously in the kitchen. Culinary sage, tarragon, and thyme are planted at corners of the outside beds, and chervil and dill seed around in the beds and gravel.

Last year, we had a serious rabbit explosion at Church House, and they found their way into the cutting garden and ate more lettuce than we did, as well as bachelor buttons, nigella, violas, zinnias, and nasturtiums. Early this spring, we will sink hardware cloth down below the fencing, in hopes of keeping them out. Bosco and I are both puzzled why we didn't do that when the fence was built. I've never had a problem with rabbits before, and remember, years ago, being amused reading an old Scottish garden book (*Flowers: A Garden Note Book*) in which the author, Sir Herbert Maxwell, ranted about the "pestilent" rabbit. I laugh no more. Last winter, they ate a prized baby specimen of a witch hazel called 'Quasimodo' almost down to the ground. This summer our 'Purple Dome' asters were sheared of all flowers the cottontails could reach, and they nipped off at ground level all the stems of a dwarf Virginia creeper we were coaxing up a stuccoed wall of the garage. Three prized plants of *Salvia* 'Argentina Skies', a special strain from the sterling garden Chanticleer in Pennsylvania, recently disappeared overnight, stem, leaves, and all. Little balls of scat left behind are the telltale sign. I spray with all sorts of organic smelly concoctions, usually after the fact, for I'm at a loss to know what the adorables will eat next. I mutter a good deal, and hope for the best.

Deer are a periodic problem at Church House, although nothing like they were at Duck Hill before we had a fence. There, a herd of twenty or more crossed through the garden at dawn every day and returned at dusk. We have no deer fence here, though I wish off and on that we did. They seldom come up to the gardens to eat in spring and summer, saving most of their destructive browsing for the woods and the wood edge. But, in the fall, they appear for dropped pears and apples, and then, most frustratingly, neatly strip the hips off the roses I've planted around the outside of the cutting garden fence. Or is it a bear that's getting those hips?

I have a weakness for roses—not the fancy hybrids that call for chemicals to remain healthy, but the wild species and their cultivars that require little care and give us a sweet-smelling show in spring before the Japanese beetles arrive, and then offer hips for late summer, autumn, and winter. I wanted some sort of plantings along the outside of the cutting garden fence, not a solid barrier, but a frame, and here, in full sun, was the perfect excuse to have some roses. I ordered rooted cuttings of *Rosa villosa*, the apple rose, through the mail, and planted them on the south and north sides of the fence. Clear pink, single blossoms in May are followed by large hips resembling tiny apples (hence its name) that develop during summer, gradually turning from green to orange to scarlet. They are outrageously decorative when they are not consumed by deer or bear.

On the north side of the fence where the soil is often wet, we have a stretch of the native swamp rose, *Rosa palustris*. It comes into bloom at the end of June, with 5-petaled, fragrant, delicate pink flowers with prominent yellow stamens. The foliage of the swamp rose is burnished with red in fall and brightened by scarlet, pealike hips. I long to add our native *R. nitida* here for its lustrous leaves that turn crimson in autumn and its plentiful deep rose-pink flowers, but it is difficult if not impossible to find.

The charming red-stemmed, hybrid *Rosa rugosa* 'Therese Bugnet' is now established on a west stretch of the fence. In May, it is lavish with double, swirled pink blooms that repeat in late summer and fall. I think it is my favorite *R. rugosa*, for it looks and smells like the true old roses but with healthier foliage. We also have the eglantine rose weeping over the fence, littered with 1-inch, coral-pink flowers fading to white centers with golden stamens. This is a large bush, its stems full of prickles, and I haven't really given it enough room here. But I love to pass close by the shrub to catch a whiff of its apple-scented leaves, a perfume that is dramatically thrown into the air on a humid day or after a rain.

Not all the roses here are pink. The first rose to bloom at Church House stands at the northwest corner of the cutting garden fence. It is the charming 'Father Hugo', *Rosa hugonis*, showering large, pale yellow flowers along its arching stems. Soon after, the burnet rose hybrid, 'Harison's Yellow', *R. ×harisonii*, begins to open its small, muddled double blooms the color of butter. I have a cherished photograph of Bosco on my dressing table in which he is standing next to a large colony of 'Harison's Yellow', I would guess 8 feet high and 10 feet wide. We discovered that rose by the deserted foundation of an old farmstead

Helianthus 'Lemon Queen' and *Rosa villosa*

in a blueberry-littered meadow while on a spring walk with friends in the hills of western Pennsylvania.

A few small-flowered clematis mingle with the roses, easily climbing the wire facing of the fence. *Clematis ×triternata* 'Rubromarginata' was a favorite at Duck Hill, and is here now lacing the fence above the apple roses. In July, it is a cloud of tiny white, 4-tepaled stars tipped with rose-purple. The annual love in a puff (*Cardiospernum halicacabum*) seeds and climbs the wire fence in summer, its green, balloonlike pods turning into papery tan lanterns in fall and winter. Up to now, I have grown fragrant sweet peas inside the garden at the foot of the fence, where they climb the wire in front of the roses and bloom in June and July. Few flowers, to my mind, can compete with a bunch of scented sweet peas in a small vase indoors.

Beyond the confines of the cutting garden and its embrace of roses is a meadow stretching out to the west and north. A mowed path leads from the cutting garden's center gate through the high grass, tempting you to leave its confines and venture out into the wild. Here, again, I like to glimpse this wildness from the geometric enclosure of the garden, to visually connect the blooms inside with the grasses and wildflowers beyond.

"My favorite gardens are the ones that have slightly gotten away from you," wrote the English gardener Claudia Rothermere in an issue of the magazine *Gardens Illustrated*. I agree with Claudia, but last summer I thought the cutting garden had gotten a little more woolly looking than I liked. It needed some architecture and repetition of plants to give it coherence. And so this spring, I marked the corners of the outer beds with small boxwoods and, in June, planted nasturtiums to edge the short ends of all the beds on the cross axis. Our young helper, John, who works for us once or twice a week, made four rustic 6-foot tuteurs this winter with cedar from our woods. They now grace the four center beds and help give a sense of order in the garden, even if annuals are seeding with abandon and rollicking about.

Scarlet runner beans on one of the new tuteurs

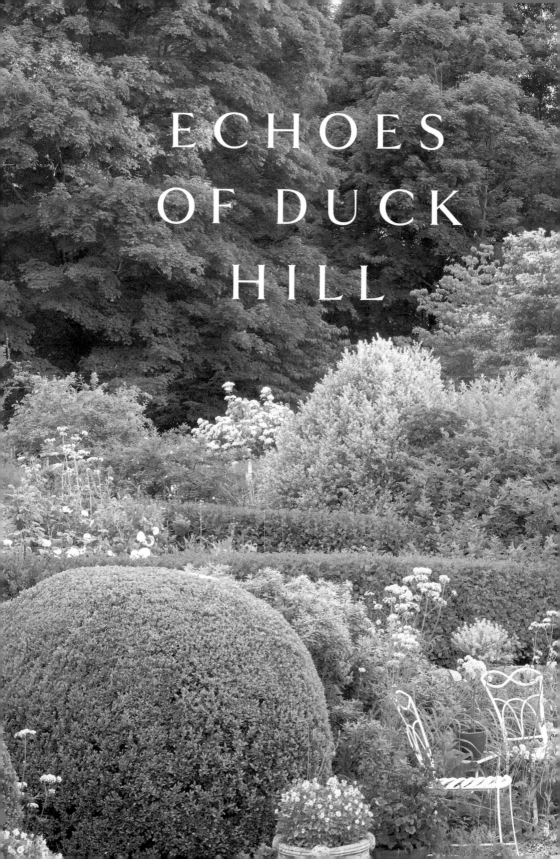

ECHOES
OF DUCK
HILL

SELF-SEEDING

Almost all my life I've had an affection for gravel—that is, for garden paths and driveways made of pea stone. I love the crunch of it, the fact that it's porous, that it doesn't cost a lot, and that it looks natural, particularly if the pea stone is small—¼ inch is good—and isn't blazing white or cold, hard gray, but an unwashed mixture of soft tans and grays from a local river source. At Duck Hill, we dressed the various terraces, courtyard, driveway, and paths with a surface of pea stone, perhaps part of the reason friends said it was like an English or French garden. Gravel is still comparatively uncommon in gardens in this country; in Europe, it is the hard surface of choice.

Nonconverts often ask me if it isn't hard to weed. No, not at all—much less tricky than pulling weeds out of brick paving, for instance, when you can easily dislodge bricks in the process. Most weeds rooted in gravel come up easily with a light tug, especially if you aren't talked into using landscape fabric beneath the stone. Bosco and I used to spend a few weeks of summer at a cottage in the French countryside that was his weekend home when he and his family lived in Paris. In refurbishing a kitchen terrace there, we mistakenly lay down the landscape fabric that is supposed to inhibit weeds under gravel. What a nightmare. Weeds happily seeded in the pea stone *on top of* the cloth, sinking their roots into the material, and when you pulled a weed up you pulled the cloth up with it. Bits of bulging cloth were soon in evidence everywhere. It's best to lay your pea stone right on top of a larger gravel used as the base in roadways, or on top of plain old subsoil as we did at Duck Hill and have now done here. The self-seeding that happens in gravel, of flowers, not weeds, turns out to be one of its most endearing qualities.

The first spring at Church House, I was struck by the absence of any flowers that had escaped their proper place to seed about. Perhaps the former owner did not allow such laxness. At Duck Hill, those

Verbascums self-seeding at Duck Hill

unexpected flowerings added much to the charm of the garden, and, because I wasn't the neatest gardener and didn't mulch heavily, all sorts of perennials, annuals, and bulbs cropped up in the gravel terraces and paths and spread in the beds. I edited out enough of those seedlings to avoid complete chaos, but left many to form pools of one flower.

Forests of tall yellow and white verbascums followed foxgloves in the herb garden there, marching along the edges of the paths and invading the beds. I loved their statuesque presence and the bird and bee life they brought to the scene. Carpets of ink-purple Johnny jump-ups colored the ground beneath shrubs and roses and often accompanied the lettuces and spinach in the vegetable garden, dribbling into the paths. Poppies wove in and out of the flower beds, the large, old-fashioned, double crepe-papery Oriental in fiery red-orange, and the tiny, seemingly fragile, apricot Atlas poppy. Shirley poppies in every hue of red to white, many double like fragile crinolines, appeared yearly in the paths and beds by the chicken house after I scattered a packet of their seed called 'Angel's Choir' there one cold, early spring. Larkspur in shades of blue flowered with equal abandon, from seed I originally brought home from Monticello. Chionodoxa, planted under a magnolia, spread into the lawn, and scilla painted puddles of blue in the herb beds. A patch of winter aconites multiplied by seed every year from a tiny cluster I brought to Duck Hill from my former home, their telltale ruffs of green appearing at distances. When in flower, they lit the woods with what seemed like sunshine, their open, glossy yellow cups attracting bees sometimes as early as February. Lavender *Crocus tommasinianus* popped up in their vicinity, uninvited by me but no less welcome.

I felt somewhat destitute that first spring in our new place without these serendipitous seedings, and went about insuring we would have them in the future. Before the new beds were even dug, large pots on the mudroom steps were filled with Johnny jump-ups, for I could not imagine a garden without their friendly faces. I knew if I didn't deadhead them, they'd manage their way into the garden. We planted generous pots of blue and purple violas on the living room terrace, hoping they would also seed about. By July, small seedlings of the Johnnies were popping up in the crevices of the stone steps in the front garden (their flowers immediately nibbled by rabbits). By the second spring, Johnnies had moved into the garden beds and clumps of blue and white violas were sprouting around the gravel terrace. Heaven.

It didn't take long. All sorts of annuals now happily appear on their own in the cutting garden. Cosmos returns in the beds and in gravel, in all its fairy shades of white, pink, and plum, single and double, without any help from me. *Zinnia* 'Aztec Sunset' and 'Persian Carpet' spill out of the beds onto the paths. I planted kiss-me-over-the-garden-gate, *Polygonum orientale*, the first spring after the cutting garden was dug, having wanted to grow this charming knotweed for years. Now the branches of nubbly, rose-pink racemes reappear on their own to dangle on their tall stems among cosmos and sunflowers. Another new self-seeding delight is bells of Ireland (*Moluccella laevis*) with 2-foot spikes of thickly clustered, pale green, outward-facing bells that are, in fact, calyces. The actual flowers, encased by the showy calyces, are white, tiny, and insignificant, and are said to be fragrant. I must remember to bury my nose in a bell next summer. This curious annual is native to the Mediterranean, not Ireland, despite its name. I suspect the Irish inference is due to its green color. Apparently another common name is lady-in-the-bathtub, although I think that title should be reserved for another self-seeder, the old-fashioned bleeding heart: for when its pendant flower is turned upside-down, and the two deep pink petals gently drawn down, they make a fanciful bathtub, and a small person is revealed half-submerged inside. The green lady's spikes cut well and are unexpected fun in kitchen bouquets.

On a March day after a snowfall that first season, I sprinkled seeds of Shirley poppies in one of the outer beds of the cutting garden, just as I once did at Duck Hill, and now they pop up every year. If you sear the cut stem ends of these fragile poppies with a flame for a second or

Shirley poppies and Johnny jump-ups in the cutting garden

two, they will last a couple of days in a vase. As the flowers fade and the stalks begin to look ratty in the garden, I pull most of them out of the bed, just leaving a few to reseed.

Chervil that I planted the first spring reseeds now in shadowy places in this garden. I don't have to plant claytonia anymore, for it appears early in the season scattered in the gravel from seed I once sowed in a row with lettuces. Also known as miner's lettuce and winter purslane, claytonia loves cold weather and has the same fleshy, succulent character as the weed, both of which I like to add to our salads. It is the first edible to appear in the garden.

I missed all the single- and double-flowered feverfew that reliably spread in the garden beds and gravel at Duck Hill, and begged a clump from Deb Munson. Within a year's time, I had more feverfew than I knew what to do with. This old-fashioned plant is a marvelous filler in the garden and in bouquets, and is easy enough to pull out if you have too much. I am fond of its nose-twisting fragrance.

When I planted the burnet roses that Katie Ridder brought as a house present, neither of us knew about the bonuses from her garden that would materialize the following spring. After the roses finished blooming, the telltale glaucous cut leaves of an opium poppy appeared all around them, and by June, 'Lauren's Grape' had opened its handsome deep-purple cups. This display was followed in summer by the tiny green bells of *Nicotiana langsdorffii*. Now, 'Lauren's Grape' has somehow leapt the wall and seeded below in the outside beds of the cutting garden, and the delicate nicotiana frolics in the terrace gravel.

Corydalis lutea and white *C. ochroleuca* spilled out of the stone walls and seeded in shadowy corners of the kitchen terrace at Duck Hill, flowering from early April until frost, taking first prize as the longest blooming perennial there. I was not without them since the first acquaintance with these charming rock plants here in Falls Village those many years ago, and longed for them again. I begged a piece of the yellow species from a friend whose nearby garden I love, and ordered a plant of the white corydalis from Digging Dog Nursery in California. Each is tucked into stone and gravel in separate places so they don't compete with each other, and, since they flowered last summer, I hope they will soon start seeding about extravagantly. I also bought a plant of cream-colored *Scabiosa ochroleuca*, the pincushion flower, to have again. At Duck Hill, this dainty perennial with deeply cut basal leaves and willowy 2-foot stems, naturalized in the graveled kitchen terrace and bloomed from midsummer through fall.

I planted the scabiosa in gravel here on a new small terrace opening out from the west side of the mudroom, with the greenhouse bordering it on the north and the kitchen pantry on its south. We are planning to extend this sheltered terrace several more feet in the spring, and I have in mind establishing plants here that like it quick-draining and are content in the strong western sun, with the hope that many will self-seed. I am inspired—unrealistically, for the scale here is modest—by Beth Chatto's famous dry garden in England and the stunning gravel garden at Chanticleer in Pennsylvania, possibly my favorite part of that superb landscape.

The pale yellow, tiny-flowered foxglove, *Digitalis lutea*, has already chosen to flower here in a shady corner by a bench, and so has a plum-colored aquilegia from seed I brought from Duck Hill and scattered here on the gravel. The columbine is already showing up in cracks and crevices on its own. Deb Munson gave me several young plants of 'Miss Wilmott's Ghost', the dramatic silvery *Eryngium giganteum*, and I planted one here with the hope that it will survive and then seed around, which Deb says it does readily. Christopher Lloyd suggests, in his book *Cuttings*, that the thistle is named after Miss Wilmott because she was pale and prickly. But the usual tale is that this formidable Edwardian lady carried seed in her pocket when she went visiting gardens and secretly sprinkled it in her wake.

I would like *Crocus tommasinianus* seeding in this terrace, just as it does at Beth Chatto's, and small poppies to mingle with feathery grasses.

Last spring I planted the 'Ladybird' poppy, *Papaver commutatum*, which I got from plantsman Helen O'Donnell. It is small-flowered, brilliantly red, with a black blotch at the base of each petal. Of course I let it go to seed with the hope that it will pepper the gravel this summer. I imagine tall mulleins here and alliums, which seeded in gravel at Duck Hill, and low asters for structure, and cushions of pinks. This diminutive playground will take years to develop with a good deal of editing.

COLD FRAMES FOR FORCING BULBS

After compost heaps, cold frames were one of the features I most wanted in the garden at Church House. We always had them at Duck Hill, in the beginning up by the barn, which was then filled with horses, hens, roosters, ducks, and geese (the barn, not the frames). When they started to fall apart years later, and the emptied barn became an apartment, Bosco and I had new frames built behind the new Boscotel and greenhouse. The purpose of these cold frames was not to grow plants from seed, though I used to do that, but to bury clay pots planted with spring bulbs for about 12 to 14 weeks starting in late October, with the joyous result of extracting them in February and March to bring inside and coax into flower.

The first year here, we bought a ready-made frame, but it didn't work very well because it wasn't deep enough. We had to dig a trench to have the depth to bury the potted bulbs. Ideally you want the cold frame to be approximately 3 feet deep in the back, sloping to a 1-foot depth in the front. That summer we asked a young friend who was a cabinetmaker with some time on his hands if he would make us three frames, 3 foot square and the right depths. When they were delivered and I received the bill, I dubbed them the Cadillac of frames, for, commensurate with their cost, they were handsomely built. We placed them on the west side of our garage, facing the afternoon sun; at Duck Hill, the frames faced south. Both exposures seem to work fine—you just want to capture the sun for several hours of the day. The frames are best tightly built with the top sash overlapping enough to ensure that no rain or snow seeps in. We learned the hard way that winter water in the frames leads to rotted bulbs.

The cold frames at Church House

On a few, preferably warm, afternoons in October, I gather the bulbs I've ordered for forcing, mostly from Brent and Becky Heath's catalog, and pick out clay pots of various sizes and shapes that are washed and ready to fill. We set up an old table on the western gravel terrace for the plantings. A wheelbarrow filled with a mixture of topsoil, compost, and grit is by my side to scoop into the pots, and a watering can with a rose for sprinkling is at the ready. I fill each pot about three-quarters full (depending on the size of the bulb), give it some water, let it drain, then place the chosen bulbs snug against each other on the surface of the soil to fill the inside circumference of the pot. A 12-inch-long wooden label with the name of the bulb, written large enough so I can read it when looking down over the frame, is thrust into the side of the pot. Next, soil gets troweled in over the bulbs, almost to the top of the pots, which are then put down on the gravel and given a second delicate watering.

The pots are now ready to go into the frames. I've used different materials to sink them into—leaves, peat moss (which makes me cough), and shavings, and find shavings the easiest to use, especially if you are in farm country, as we are, where an Agway will have bales of them. Once you have the shavings, you can use them over and over if you don't get them wet. The bottoms of the frames, which are open and sit on the ground, are filled with about an inch of gravel, then several inches of shavings on which the pots will stand. When all the pots are placed, shavings are scooped in and around, and piled on top, sometimes as high as the sides of the frame permit. I tend to put the littlest pots of bulbs toward the front of the frames where the depth is 1 foot, and the biggest pots at the very back and deepest part. Once the pots are all in and snuggly covered, you hope with their labels peeking out, the top of the frame is closed.

Before Duck Hill, I lived in an Edwardian house where a series of old cold frames existed by an old shed behind the kitchen. I used those frames to grow perennials from seed, inspired by the gardener and author Louise Beebe Wilder, for a long time my guru. But I found that the panes in those sashes shattered with the least impact if they accidently banged closed or were hit by a wayward branch. Discarded glass storm windows were used successfully as sashes on the first frames I had at Duck Hill. But now we have stiff, clear plexiglass surrounded by wood as the sash, which is lighter to lift and doesn't tend to break. The back of the sash is hinged, and because our frames are just a few inches from the garage wall, the sash can lean back against the wall while I am putting in or removing pots. Otherwise you can prop a sash open with a long stick, or a brick, which I sometimes do when the temperature hovers around 50°F. You don't want the bulbs to cook. Or to freeze.

All sorts of little bulbs are perfect for forcing in pots and, if you've more than one pot of a variety, they make delightful presents to give to friends when in bloom. *Scilla siberica* 'Alba' is one of my favorites—it's a little more unusual than the deep-blue form that colors lawns and garden beds—and its sprays of small, starry flowers have a delicate, loose, dancing grace that I love. Four-inch pots of this squill are nice to have on a table by your side. The white form of glory-of-the-snow, *Chionodoxa luciliae* 'Alba', is stiffer stemmed, with bigger, out-facing stars, and is good-looking in a 6-inch pot. Sometimes, for variation, I pot up some *Puschkinia scilloides* var. *libanotica,* which is similar in flower shape and habit to scilla, but has milk-blue petals with midveins of dark blue. It is commonly called striped squill.

Grape hyacinths, varieties of *Muscari*, are perfect for forcing, and now we often see the common blue *M. aucheri* potted up for sale at supermarkets near Eastertime. I always have some pots of white grape hyacinths (*M. aucheri* 'White Magic') in the frames, for this is another favorite. But each year I try one of the more uncommon sorts, like *M. neglectum* with stalks of darkest purple florets edged in white. Dutch crocuses, of course, are another spring bulb we can now readily buy pots of in late winter. I'm more inclined to force the small-flowered sorts, varieties of *Crocus chrysanthus,* such as the white and purple-blotched 'Ladykiller' and pale 'Cream Beauty' with deep orange stamens.

It is fun to force more unusual bulbs that you don't see in a super-market. Fritillaries do extremely well in cold frames, especially the white and checkered bells of *Fritillaria meleagris*, or the dusty purple, miniature lampshades of *F. uva-vulpis*, tipped in yellow. What could be more charming in a pot? Maybe trout lilies, also called dog-tooth violets, varieties of *Erythronium*. We force the elegant yellow hybrid 'Pagoda' and the white *E. californicum* 'White Beauty' in 6- or 8-inch pots and are rewarded with 8-inch stems above mottled green leaves, each supporting 3 to 5 exquisite, miniature nodding lilies.

I force more daffodils than any other bulb, probably because it is one of my most-loved flowers. (I dread when asked at a lecture what my favorite flower is, for I can't possibly narrow the field to one or even two. But narcissus would hover somewhere near the top of a list.) I mostly choose the smaller flowered varieties to force, but this is my preference; trumpets and large-cup narcissus can force well and look splashy in big pots.

The jaunty cyclamineus types with flared-back petals (or segments of a perianth), are aptly described by the Heaths as "looking like they are standing in front of a fan." They force easily and, perhaps because most are early blooming in the garden, are quick to bloom in a pot and are the first daffodils I pull out of the frames. 'Jack Snipe' is one I invariably include, 10 inches tall with white swept-back petals and a butter-yellow cup (also called a corona). Bosco likes jazzy 'Jetfire' for its bright yellow, reflexed petals and red-orange cup. I fell for another cyclamineus narcissus years ago at the San Francisco Flower Show confusingly called 'Snipe'. It is a delightful small thing with narrow, flared, cream petals and a very long, pale yellow, cylindrical trumpet. For some reason, it is sometimes difficult to find in catalogs and is expensive when you do, so I content myself with a couple of small pots of five bulbs each.

The miniature daffodils are sweet in pots—sorts like the tiny golden yellow trumpet 'Little Gem', and 'Toto', 4 inches tall with several white and cream flowers to a stem. 'Rip van Winkle' is always on my list to force, its dandelion head of slivered yellow petals unexpected, and certainly more comical than comely. 'Minnow', on the other hand, is downright pretty in a pot, a diminutive tazetta narcissus with the clusters of flowers typical of this division, white perianths around tiny, pale yellow cups, on 6- to 8-inch stems. The triandrus 'Hawera' is equally charming with clusters of pale yellow bells nodding on 6-inch stems. For elegance, 'Segovia' is a daffodil I treasure, in the flower garden or in pots. It is delicate in stature, with overlapping, rounded white petals and a flat, crimped, light yellow cup. Forced bulbs tend to grow a little taller in a pot than they do outdoors in the ground.

The species tulips are beguiling to have indoors, opening their cheerful cups in a low, wide clay pot in late March or early April. Try *Tulipa batalinii*, merely 4 inches high, in bright yellow or bronzy apricot; and the little lady tulip, *T. clusiana*, affectionately called the radish tulip, its pointed white petals flashed on the outside with red. The flowers open up, starlike, in the sun to reveal interiors that are all white with a small purple blotch at their base. 'Lady Jane' is a taller, bigger flowered hybrid of the species, its candy-cane blooms on willowy 12-inch stems.

Many of our friends don't have cold frames but force bulbs anyway. You can do it in your cellar if you have one of those outdoor hatches, placing them on the bottom step (and, if they are crocuses or tulips, protecting them from mice). You can put them in a Styrofoam box, the sort you might use for a picnic, and have them in a garage, if it's not utterly freezing there. Or, if you have an extra refrigerator, you can stash the pots in there. The idea is to give them a three- to four-month chilling period without letting them freeze.

It is late February, as I write this, and I am now weekly plunging my hands in the frames to pull the first of the bulbs—pots of hyacinths, crocuses, scilla, the earliest daffodils. I carefully dust off the shavings from around the fresh tips of growth and give the pots a good drink of water. Then I place them by a window or take them into the greenhouse to warm up and slowly grow and begin to bud. The closer we are to spring, the faster this happens.

Bulbs blooming inside in winter seem to me a delightful alternative to florist flowers, and a lot cheaper, if you don't take into account the cost of the Cadillac frames. Longer lasting, too. We like having a pot of some sweet bulbs on the kitchen table where we eat our meals, and another in the living room on a side table. If we have a particularly splashy pot of blooms—trout lilies, or daffodils, or tulips—we put it in the middle of the long table in the mudroom, so that the spring flowers greet us as we come indoors in our parkas and boots.

These forced bulbs don't go to waste after blooming. Not with Bosco, my refugee husband, around. I used to toss bulbs on the compost heap when they were spent, just as we do with tender paperwhites. Now, we keep them watered after their flowers are over, storing them in an out-of-the-way place, for their foliage soon looks a mess. Then, either in late spring, when the unsightly leaves can finally be cut down, or in fall, when we hope we can still read their labels, we dig them into the garden beds or the orchard meadow. At the start of our marriage, when Bosco insisted on saving and replanting the bulbs, I was extremely dubious about them flowering again. To my amazement, they didn't skip a beat, blooming the next spring. Now, when ordering bulbs for the cold frame, I not only consider what varieties are nice in a winter pot, but how they will subsequently dress the garden or orchard.

A SMALL
GREENHOUSE
FOR BOSCO

As part of our determination to scale back our gardening and our expenses, to simplify, Bosco said he would not have a greenhouse again. But he missed it terribly. The greenhouse at Duck Hill extended out from his one-room Boscotel hideaway, and gave him hours of immense pleasure in winter and early spring. There, listening to opera on the radio, he sowed his heirloom tomatoes, growing over a hundred pots of twenty-five varieties so he could sell them at the spring library plant sale. There he had his calamondin orange from which he made a killer jam. The 12-by-20-foot space—two salvaged bays of an old Lord and Burnham glass structure—was cheerful with flowering and scented geraniums and the African bulbs we loved to grow, lachenalias, veltheimias and babianas, as well as fragrant freesias. Bosco started begonias from leaf cuttings in shaded corners, and raised assorted annuals from seed, potting and repotting, loving his warm winter haven.

During our first winter at Church House, he started a new collection of begonias, begging leaves from friends and coming home from nurseries with just one more strangely spotted and crinkled variety. We enjoyed geraniums in the bright sun of the living room and on south-facing windowsills in the bedrooms, and contented ourselves with ferns, begonias, and clivias in the mudroom, where, to our dismay, the low winter sun from the west was mostly blocked by our new addition.

But Bosco dreamed of a greenhouse. By the second winter, he was poring through catalogs and looking up glasshouses on line. The completed renovation of the house, of course, was more costly than expected. We knew it was prudent to tighten our purse strings, and having a greenhouse was a luxury. But then we thought of the delight for Bosco such an addition would provide. We decided to plunge—with the goal, this time, of building a simpler, smaller, more economical structure.

We were intent on having a greenhouse this time that was energy-efficient (ours at Duck Hill was not), and, to this end, we engaged the help of our neighbor Bill Devries who possessed a wealth of knowledge about glasshouses and hoop houses and how to build them sustainably. Bill was the past owner of a large construction company, which seemed to me incongruous, for he had an air of fragility about him, a quiet, soft-spoken, thoughtful man, tall and thin, dressed in plaid woolens, with wisps of hair around his ears and a long white beard. He became our partner in crime.

Bosco discarded the idea of a freestanding structure, preferring it to be attached in some way to our house, for he didn't want to be forced to put on his boots and snow jacket to get there. At Duck Hill, if he was down at the house rather than at the Boscotel, he had to do just that, climbing up stone steps through the garden, brushing past wet boxwoods on a rainy or snowy day, then crossing our tiny orchard to reach the outside door.

Here, the western wall of our old stuccoed garage was the only viable spot where we could break through and add a greenhouse; and Bosco argued that it was convenient, for the garage was connected to the mudroom, and he merely had to open a door and walk down two steps to get there. I shuddered at the idea of having to pass by garbage cans and the inevitable mess of our garage to reach the greenhouse, but I figured a narrow space could be carved out and eventually fashioned into an attractive potting area as well as a throughway to the green-house, with enough room left for a car. Somehow, with shelves or a partial wall, we could shield the route there from the main part of the garage where all the piles of boxes and newspapers and garbage waited for their eventual journey to the dump.

The new greenhouse

We wanted Bill to build us the greenhouse from scratch, but he was retired and busy enough without taking on such a task. He agreed to assemble one, if we bought the parts. Bosco admired Katie Ridder's small, good-looking greenhouse attached to her garage at one end of her enchanting flower garden in Millbrook, New York. Her structure was not wildly expensive and came in precut and labeled parts from Arcadia Glasshouse, an outfit in Madison, Ohio, recommended to her by the New York Botanical Garden. With Bill's approval, Bosco chose a structure from Arcadia that was 10 by 16 feet and would jut out from the garage with an attractive peaked roof. The side walls would be $\frac{3}{16}$-inch-thick glass, and the roof would be made of 16-millimeter polycarbonate, which looks to me like smoky, ribbed glass, a material that provides added insulation and, as a bonus, prevents the glare of overhead glass.

In October 2017, we began to excavate. Our friends Deb Munson and Naomi Blumenthal, both professional horticulturists with a lot of experience working in greenhouses, suggested we attempt a partial pit house, digging down far enough that only the glass sides of the house were above ground. We loved the idea, but were foiled in the end, because we were only able to dig down half the depth because of ledge rock. Blasting was not an option. Bill convinced us to use insulated forms called Logix as our footers for the structure rather than the more traditional brick we had at Duck Hill. They were like Lego pieces fitted together, made of Styrofoam into which concrete was poured, and were a cheaper solution as well as providing excellent insulation. I was horrified at first by their ugliness, but, after the greenhouse was built, we were able to cover the footing with stucco and paint it white, echoing the façade of the garage.

Surprises are inevitable when building and rarely heart-warming. Mystery electric wires, a telephone wire, and water pipe were uncovered as we dug for the greenhouse foundation, resulting in days of unexpected work by an electrician and plumber. When we broke through the west wall of the old garage for the connecting door, we discovered that it had no foundation left (made with cinder blocks when they were still cinder, said the amused mason.) The wall was about to collapse and needed to be rebuilt. We groaned. The work proceeded.

All the pieces of the greenhouse that Bosco ordered lay on the lawn, each length of white metal numbered, with a handbook to help with the assembly. For the next several months, Bill worked by himself with his ladder and drills, putting it all together. I watched through the kitchen windows as the greenhouse slowly rose from the ground. In January, the electrician and plumber came and installed propane heat, lights, water, and a fan. Simple metal benches were ordered online and installed along the walls and in the center of the space. By February, Bosco had his working greenhouse. The happy acquisition of plants ensued, with a trip to the irresistible world of Logee's nursery in Danielson, Connecticut, returning home laden with citrus, jasmines, violets, and cyrtanthus in flower. Presents of zonal and scented geraniums, billbergia (commonly known as queen's tears), ferns, and succulents appeared from generous gardening friends. I started to pull pots of bulbs out of the cold frames to coax them into flower on the benches. Very quickly and happily the greenhouse began to fill up, a joyful sight.

Looking back, we regret a decision or two. Isn't this always the case? In the interest of retaining heat, we agreed to have a cement floor; but it looks a mess most of the time, showing every speck of dirt, and we wish we had chosen a more attractive flooring, maybe gravel under the benches, which would hide any dirt, and a pattern of wood divisions crisscrossing and framing the concrete along the paths. We also wish the greenhouse had a door to the outside on its far end, but the model Bosco chose only came with a single opening, the door attached to the garage.

This winter, we had our first setback. On a night of single-digit cold, our greenhouse heater shut down because of a dirty filter, and the next morning we were faced with a massacre. Only the rosemaries

Inside the greenhouse

survived. Bosco was practically in tears. We slowly cleaned up the mess, Bosco emptying and washing countless pots, and we started again. It is early March now, and our greenhouse is brimming once again, mostly with gifts from sympathetic friends. We look out complacently from our kitchen windows across a snow-laden terrace to misted glass enclosing our cossetted jungle of green foliage and flowers. Their well-being is checked many times a day. I am campaigning for a heat alarm attached to my phone.

CRAVING WOODIES

It is early May as I write this, cool, gray, and wet outdoors, and the shadblows that weave among the evergreen spruces at our southern border are sparkling in the grayness, their clusters of tiny, fragile, 4-petaled flowers a blizzard of white, though the young accompanying leaves are a soft russet, and the color combination is unusual. In a day or two their flowering will be over, but, even as the white petals drop to the ground like so much confetti and the leaves shade to green, the remaining stems and seedheads, tinted a soft pink, will seem like a second flowering. By June, the developing fruits will bring birds to the garden, cedar waxwings and robins, and, if we're lucky, grosbeaks and thrushes. Shads can be glimpsed delicately flowering in woodland now as we drive around the neighborhood, and I feel incredibly fortunate to have a wealth of them in the garden here.

Even as the shads are flowering, two old bushes of *Viburnum carlesii* by our garage are opening ball-like clusters of pink-flushed white blooms, and the air all around is filled with their perfume. My guess is that Nancy McCabe planted these viburnums along with the shadblows, crabapples, and kousa dogwoods at the garden's edge. But almost no other shrubs besides a few lilacs were in evidence in the cultivated areas around the house when we moved in. I missed the rich variety of shrubs we grew at Duck Hill, realizing how important these woody plants continue to be for me.

When I first started seriously gardening in my late-twenties and early thirties, I was enthralled with perennials, voraciously reading and rereading books on the subject by my mentors Louise Beebe Wilder, Graham Stuart Thomas, and Fred McGourty. I collected what I could find and what I could afford. White Flower Farm was the only mail-order catalog I knew in those early days that offered a variety of plants, and every year I ordered one of this and one of that. The perennials that

The fruit of *Viburnum opulus* var. *americanum*

grew lustily for me were periodically divided and replanted until I had satisfying patches and sweeps of each.

But by age forty, I was engrossed in a new passion—it's what happens with gardeners. This time it was old shrub roses, spurred on, again, by the writings of Graham S. Thomas, who famously brought these antiques out of obscurity. They virtually disappeared from gardens for approximately fifty years after the stiff but sexy, ever-blooming hybrid tea came on the scene in 1900. Thomas searched out the all-but-forgotten roses in a few private gardens in England and France where they were still to be found, began to propagate them, and wrote a seminal book about them in 1955 called *Old Shrub Roses*. A great revival of interest began. Three decades later, I pored over the descriptions of these fragrant, lavish shrubs, many of ancient lineage, in his book, and in essays by Vita Sackville-West, who, inspired by Thomas, passionately collected them for her garden at Sissinghurst. I wasn't the only beguiled American. By the '80s, interest was stirred up in the States, and a few stalwart gardeners and rosarians rustled for these roses in old farmyards and cemeteries, took cuttings that would save and perpetuate them, and, as near as possible, recorded their identities.

I ordered them bare-root from Tillotson's (eventually Roses of Yesterday and Today) in California and Pickering's in Canada, mail-order nurseries sadly no longer in existence. (Why do so few nurseries in the United States today offer old and species roses? Am I one of a very few who still wants to grow them?) A carload of *Rosa* species—albas, centifolias, damasks, and gallicas—came with me when I moved to Duck Hill in 1981. These became the bones of the first gardens I dug here—the main garden and the herb garden—and lent them an unrivaled air of romance. Over the years, a number of wild species roses were added to the mix as I became increasingly interested in roses impervious to bugs and disease that would flourish in an organic garden.

My interest broadened to all kinds of shrubs by the time I was in my fifties and sixties, entranced, finally, with all they offer in structure, flowers, fruit, and fragrance. The appreciation of shrubs tends to come to us late in our gardening life, I think. Again, I repeatedly dipped into books to learn and research—volumes of wisdom from the Arnold Arboretum's Donald Wyman and from Michael Dirr in his *Manual of Woody Landscape Plants*. I read with delight the poetic observations of the plantsman William Cullina in his *Native Trees, Shrubs, and Vines*, a book I still return to every spring. I studied the yearly catalog from Broken Arrow Nursery in Hamden, Connecticut, known for its stellar selection of shrubs and small trees, and marked in red ink with must-haves. And so, Duck Hill became rich in shrubs that, over the years, reached a magnificent maturity, their height and width always greater than I expected. They brought birds and butterflies, threw fragrance into the air, and offered lavish lengths of branches to cut for indoor displays. When we moved here, I longed for some old favorites again.

Well, I've managed, almost without noticing, to accumulate quite a number in four short years. The trick, I repeatedly have to remind myself as I bring home one more woody plant, is not to clutter, to be careful at Church House not to do what I did at my old garden, dotting shrubs and trees over every available space, destroying all the quiet of the place. The large front meadow here is inviolate, and I am loathe to add to our woods any plant that is not a native already found there. Shrubs and trees indigenous to our area of New England that I think will thrive here are slowly being added to the wood's edge in view of the house, enriching our habitat for birds. But, newly bought, non-native shrubs, "gardenesque" sorts, are confined to the garden beds, or close to buildings, or near our conifers and shads along the property line.

Like a brand new gardener, I'm impatient for the ones we've planted to settle in and start stretching their limbs. I might never see slow growers like the March-blooming witch hazels relax their branches and knit together the way they did at Duck Hill. But, in ten years' time, I'm bound to witness some of the small shrubs we've planted grow to a respectable size. That old maxim, preaching patience toward newly planted shrubs or perennials, comes to mind: "Sleep, creep, leap."

Fragrant abelia (*Abelia mosanensis*) was a gardenesque shrub I loved at Duck Hill for its beauty from spring through fall, planted there in a mixed border of shrubs, perennials, and roses. Here I've planted it again along the foot of our south-facing terrace wall, where we can catch whiffs of its sweet perfume. If Sadie will stop digging it up in her pursuit of voles, it ought to start doing well. Some websites say it likes an acid soil, and the soil at Duck Hill was slightly acid, enriched over the years with manure and compost. Whereas here, the existing soil near the house, sitting on limestone, is alkaline. The abelia is planted in topsoil brought in after the construction; I need to test it. But I might have to mulch the bushes with white pine needles to make them feel at home. The flowers of this unusually hardy Korean abelia are white, 5-petaled pinwheels, opening from long, tubular, rich pink buds, the overall effect of the shrub seeming prettily speckled white and pink. Flowering occurs in early June and then sporadically in September on the arching stems, and the glossy foliage turns beautiful shades of orange and red in the fall.

So do the handsome, dark green leaves of *Viburnum ×burkwoodii* 'Mohawk', which we had at Duck Hill outside the dining room windows. Its habit is taller and more graceful than *V. carlesii*, one of its parents, though its fragrance is not quite as intoxicating. A small specimen of the burkwood viburnum now flowers in mid-May at the edge of the western terrace against our pantry wall.

We've planted a line of our native cranberry bush, *Viburnum opulus* var. *americanum* (what used to be called *V. trilobum*), to shield a parking space off the driveway. It's not a hedge exactly, because it won't be clipped and has a loose, natural way of growing to about 8 feet. The maplelike leaves of this viburnum are attractive in spring, as are the white lacecaps. But best of all are the decorative clusters of red, translucent fruit that droop from the branches through fall and most of winter. At Duck Hill, I grew it as a light screen along the road and loved how cedar waxwings came in a flock to consume the drupes sometime in late winter.

The snowballs of some doublefile viburnums (*Viburnum plicatum*) are decidedly gardenesque; there's nothing wild-looking about them. The selection 'Mary Milton' made me weak in the knees the first time I saw it in flower, with fat balls that were not the typical white but a pale, dusty pink resting against bronzed new leaves. We had a bush of it behind the barn at Duck Hill, and when we left, it had just reached a good enough size to justify cutting branches for a bouquet. We now have a small specimen by the compost heaps, and it bloomed for the first time last spring—opening not pink, but white. What? Maybe it will think better of its ways and perform in pink this spring.

When I consider a shrub to add to our garden, I think about its beauty in the landscape and its appeal in more than one season. If it's native, of course, I value its attraction for bees and birds, as well as the larvae that become our butterflies, which are essential food for newborn birds. But, whether I'm choosing a Japanese snowball viburnum or an American elderberry, inevitably I wonder how it will be to cut for the house.

At Duck Hill, I always had a bouquet in a fat white pitcher behind our sofa in the library, and the basis of that arrangement was invariably branches of some shrub. Our ceilings were low, and the room was small and cluttered with books and bibelots, so there was a limit to how tall and wide those branches could be. In spring, I used delicate sorts like Virginia sweetspire (*Itea virginica*), its graceful, fuzzy white racemes combined sometimes with stalks of variegated Solomon's seal, and lacy ferns. In August, I cut the nubby white flower stalks of clethra and mixed them with fat heads of *Hydrangea arborescens* 'Annabelle'. Later, the conical heads of peegee, or panicled hydrangeas (*H. paniculata* 'Limelight' and 'Tardiva') just turning from white to pink, would be gathered and mixed with branches of *Viburnum dilatatum*, sporting its

Itea virginica in autumn at Church House

cymes of tiny red fruit. It was a weekly challenge I enjoyed to cut new branches of some shrub and mix them with seasonal flowers, leaves, and vines. Now at Church House, I gladly continue the custom behind our living room sofa, and, with the high ceiling, I can play with taller, more dramatic displays.

Virginia sweetspire is not native to New England, but to swamps and wet woods in New Jersey and Pennsylvania and down the East Coast. Its cultivar, 'Henry's Garnet', so far survives in our Zone 5 climate, and in bloom it is a delight in the garden, its "kitten-tail spikes," as William Cullina describes them in his book, *Native Trees, Shrubs, and Vines*, arching away from the mound of foliage "like trailing fireworks." The nectar of those flowers is a magnet for butterflies, and in autumn the long, lustrous leaves turn scarlet, burgundy, and orange. At Duck Hill, a colony of Virginia sweetspire was mingled with fothergilla in a moist, partly sunny border, the two of them outdoing each other in autumn display. Now I have it in a damp, sunny corner by the cutting garden where I hope it will sucker and spread enough to justify cutting a few stems to bring indoors.

Fothergilla is here now, too, the hybrid variety 'Mount Airy', planted in the shade of our big sugar maple by the house, and there is not a moment, to my mind, when this southeast native is not beautiful. It has a wavy, zigzag twigginess that appeals even in winter; its licorice-scented bristlelike blooms are charming, progressing from a lime-green to white in May; and the handsome ribbed leaves are aflame in the fall.

Summersweet, *Clethra alnifolia*, is native to wet woods up and down the East Coast, including New England. It flowers for us in August, sweet-smelling white spires rich in nectar, alive then with butterflies and bees. I once walked a path among huge, mounded colonies of naturally occurring clethra that filtered through a low, wet wood in Bedford, New York, and I remember being astonished by the fragrance and late flowering in that shadowy, moist place. (The owners of that wood, dear old gardeners—oh, probably my age—had made a path through the wet by sinking telephone directories, when we had an abundance of such things, that were quickly covered by leaves and moss.) Even after the flowers are shed, clethra's beaded, upright seed heads are decorative, tinged with red when lit by the afternoon autumn sun, finally deepening to chocolate brown by winter. A true, all-season shrub, its leaves turn a rich golden-yellow in fall. Clethra suckered and colonized at Duck Hill along the woodland path and behind the herb garden, at home in shade and in sun. I hope it will flourish here, though, again, it is said to prefer an acid soil. I've dug in the species in a wet partially shaded clearing near the pool garden; and several bushes of its fancy pink form, 'Ruby Spice', which Broken Arrow Nursery introduced to the trade, are clustered below the kitchen window. We have a spigot there for filling watering cans, and the summersweet benefits from the inevitable spillage.

Near our southern property line, we have an old bank barn, dating, we think, from sometime in the nineteenth century. Its vertical boards were once painted burgundy-red, and are now faded and streaked with gray where bare wood shows through the bleached paint on all but its north side. We treasure the barn mostly as the southern background view from our bedroom window and from the living room terrace and the cutting garden; and we love it also for its history, imagining the farm life that once was here. Bank barns were traditionally set into a hillside, so that they could have openings and uses on several levels. Cow stalls still remain in the lower level of our barn next to a large, open area we are told was possibly for pigs or for sheep. In the vast upper level, where slivers of light stream through the old siding, hay is no longer stacked. Rather, we unceremoniously store our garden furniture here in winter and the largest pots from the garden.

I felt justified in having a few non-native shrubs by the barn. Bosco planted some rows of raspberries, currants, and gooseberries last spring, down one side of the bank where a door opens to the cow stalls. The ground is level here, shored up by boulders as the bank dips even lower, and we imagine it was used as a small yard for animals, maybe where the cows could mill about and get some sunshine. A few large stone steps descend to a lower field and the back of the barn. Here, looking up, its full height is revealed, for there is a dramatic drop in elevation, rising to a peaked roof from a foundation of random, piled rocks. The field here is quite moist, rather rough, the soil rich from decades of animal droppings. I've planted some pussy willows here—not our native *Salix discolor*, which we have in our small wetland, but the black pussy willow from Japan, *S. gracilistyla* 'Melanostachys', with deep red stems. We had the black pussy willow at Duck Hill in the yellow garden near the house, and the small, bristly catkins opened in February, deep red at first, then black, and finally, as the anthers appeared, speckled yellow. The bush grew to about 8 feet, rather upright in habit and graceful. But maybe the grace was helped by my raiding it for vases indoors, which caused the stiff stems to multiply with thinner, more relaxed branches. A year before we left, a little yellow-bellied sapsucker busied himself over the winter making so many circles of holes in the trunk, the shrub died down to the ground. I am certain it would have revived itself.

I added two other willows to my order this winter, certainly not because I need them, but because I can't always help myself when I am cruising through a good nursery list on the internet. One is a fairly new cultivar of the rose-gold pussy willow, *Salix gracilistyla*, called 'Mount Aso', with luscious fat pink catkins opening in February and March. It is said to grow to a moderate size, about 6 feet tall and wide, and was developed for the cut-flower trade in Japan. With its smaller size in mind, I've planted it in the pool garden, which is fenced and thus out of bounds for deer. The other is *S. ×erythroflexuosa*, the scarlet curly willow. Coppery red twirled stems and yellow catkins, a flower arranger's curiosity. How can I resist? It now resides by those great stone steps at the foot of the barn field. The salix came from Michael Dodge's amazing Vermont Willow Nursery and arrived by mail in April in bundles of five long straight budded sticks. Directions say to dig holes with a crowbar and plunge each stick a remarkable 8 inches into the ground. This leaves about 2 inches of willow above the soil, at this point looking pretty silly, and I am waiting to see them sprout. Since I don't need five of any willow, assuming they do sprout, I have given some of them away to a friend with a wet hedgerow. Until they get really big (how many years?), we need to protect the willows against deer browse and so have encased the little sticks in cylinders of wire. As it happens, bears are a consideration too. We've learned that willows are our local black bears' favored food when they first come out of hibernation. I hope they are content with the native willows in our small wetland and the vast wetlands nearby.

At the barn's highest elevation, on one side of its great sliding door, we planted a small specimen of beauty bush, *Kolkwitzia amabilis*, that we stumbled across at a plant sale at Berkshire Botanical Garden. This old-fashioned shrub, which is native to China, is almost as ubiquitous as lilacs in northwest Connecticut, standing at the corners of houses and by churches. Its arching branches of pale pink, bell-shaped, weigelalike blooms are a gorgeous sight in May. It was the only shrub we inherited at Duck Hill besides lilacs, and we left it where it was growing in back of the herb garden next to an arbor we built of locust from the property. The bark of the beauty bush is pale and shaggy, and attractive pink-tinged, hairy calyces linger after the splash of spring flowering. But there is no vivid fruit or autumn coloring, and the blooms are scentless, which seems an insult. I'm wondering now why we planted it here. Michael Dirr, in his seminal and always opinionated book on woody plants, calls the beauty bush coarse and cumbersome, and he says, when out of flower, it gives him a headache. I wouldn't go so far as to say that, but we did buy it in a rash moment, out of sentimentality (how often do we gardeners do that?), squandering a spot, I now rather think, that could have gone to a worthier subject. At Duck Hill, I let the rambling rose 'Trier', which scrambled about on the arbor, also thread through the beauty bush, which prolonged its appeal.

The old bank barn at Church House

From the bank front of the barn, we have a rough track to the road through a swath of high grass, on the south side abutting our neighbor's lawn. When we first moved in, I thought how romantic it might be to have a hedgerow of lilacs here as a screen, knowing these super-hardy shrubs would thrive in the full sun and our alkaline soil. We already have two lilac bushes (cultivars of *Syringa vulgaris*) at the front corners of the house, one a deep maroon, another white-flowering—I suspect, both planted by Nancy McCabe. An old raggedy stand near the garage of the ordinary lavender doubtless predates her. You would think that would be enough really, three different colors. But I was nonetheless tempted to collect other selections and colors, double white, pink, blue, and ink-purple, as well as one of the early flowering *S. hyacinthiflora* and a variety of the late *S. prestoniae* to extend the brief period of flowering. A hedgerow in full sun seemed ideal.

But the more I thought this spring about having a whole border of lilacs, as luscious as it would be in flower, the more it weighed on me that I would be planting in multiples a shrub that, like the beauty bush, had little to offer after its few days of bloom. The foliage inevitably gets mildewed, the shrubs themselves, unless of a venerable age, are rather gangly, and no autumn color or fruit redeems them. And nothing to attract bees or butterflies or birds. Surely I could plant something with more to offer, a plant that would enrich our wild habitat?

I've now decided as a compromise to include stretches of our native nannyberry, *Viburnum lentago*, in this screen of shrubs, along with some lilacs. Nannyberry grows 8 to 15 feet in height with a wide spread, has beautifully ribbed, lustrous leaves, arranged, as with all viburnums, in opposite pairs on reddish brown twigs, and has good red fall color. The cymes of small cream flowers in May are followed by edible blue-black fruit that ought to attract birds. It is also called sheepberry, possibly because "the ripe and rotting fruit smell like wet sheep wool," according to the University of Connecticut College of Agriculture. I'm not altogether sure what wet sheep smells like.

I am unlikely to ever give up lilacs any more than I'll forgo peonies or daffodils. Our common lilac is so indelibly associated with old farmhouses in New England and northern New York, you would think it a native. But, in fact, it is native to southeastern Europe, and was brought here by the earliest settlers to the colonies. The lilac specialist, Father Fiala, wrote in his seminal 1988 book, *Lilacs: The Genus Syringa*, that the oldest living specimens, dating from the 1750s, still thrived on Mackinac Island in Michigan (first brought there by the Hubbard family from New Hampshire who came to the island to farm), and in Portsmouth, New Hampshire, "massive old trees, gnarled and twisted with the centuries." Both Mackinac Island and Portsmouth hold lilac festivals in June. I wonder if those ancient specimens are still there.

Lilacs bloom on new wood produced the season before, just after flowering. So any pruning you do in early spring will forfeit this year's flowers, and pruning you do in the fall will forfeit next spring's bloom. Of course, if you feel the bush would benefit from a hard renewal by cutting an old branch or two down to the ground, or there's dead wood to remove, this can be done at any time. My favorite time to prune lilacs is when they are in full flower by cutting branches to bring indoors for vases or give to friends. I am careful to cut just above a bud, for that's where new flowering growth will happen. The cut lilacs don't last many days in water, sometimes as few as two, even if you strip off the leaves and add a drop of Clorox to the water or an aspirin; but, oh my, how luscious and sweet-smelling they are in a room, however briefly. Sometimes, after the lilacs finish blooming outdoors, I cut off the faded flowers for the sake of neatness. But I often don't get around to it. And the dark winter seed heads aren't bad to see on a snowy day. Whether lilacs are deadheaded or not has no effect on their flowering. Think of the great old stands we see when driving around in spring that are never deadheaded because you would need a tall ladder, and who's going to do that? They manage to put on a fabulous display anyway. Like apples and crabapples, the flowering tends to be most lavish every other year.

PLANTING AN ORCHARD

When we began searching for a new home, I dreamed of an apple orchard as part of the package, a pleasing pattern of mature, rounded, rough-barked trees, stretching their limbs above a high-grass meadow, and in the best of worlds that meadow clotted in spring with daffodils. The old church we chose didn't have that orchard, but two ancient picturesque apple trees guarded the house, and a third leaned on its side at the edge of the quarry, its branches gray-green with crinkled lichen, miraculously still producing small, delicious, McIntosh-type apples every other year.

The previous owners of Church House left an old rusted steel fire pit at a high vantage point behind the house. We sit there now on teak benches silver with age, sometimes warmed by a fire, admiring the splendid view of the Berkshire Hills and our property spread out below us. From the fire pit, a field slopes down to the south ending at the new cutting garden, which juts out from the living room. One of the venerable apple trees is at the edge of this field, and two old pear trees, not great beauties but carrying the weight of history, stand some paces away. This seemed the obvious spot to have a small orchard, somehow knitting the new trees in with the existing pears and apple. I imagined sitting by the fire pit looking down on a sea of apple blossoms in the spring.

At Duck Hill, we had a tiny, productive gathering of apple trees on a slope of grass between the greenhouse and our entrance walk. Early on, I planted semidwarf whips of old-time apples: Gravenstein and Cox's Orange Pippin, Mutzu, and Empire. At some point a Granny Smith was added, I'm not sure why, and a Honeycrisp. A friend gave us a whip of a quince tree that, in a surprising few years, provided the sweetest pink flowers and fruit for compotes and paste. The Cox's Orange Pippin grew up enough to give us a couple of years of delectable apples before mysteriously keeling over one spring. A critter had eaten all its roots. All the other apples, however, were substantial trees by the time we left. Their fruit, organic and thus imperfect in appearance, gave us a wealth of good eating for many years, out of hand and in sauces and pies.

This time, I decided against planting whips, thinking we needed a jump-start on their eventual fruiting. I bought 4- to 5-year-old trees, which seemed expensive enough, mostly semidwarfs, and planted them in a grid, 30 feet apart from each other, allowing for how big mature apple trees get, even semidwarfs. But the new trees looked little and forlorn in the field; you could barely see them. Bosco gazed at them and then at me and said "Do you know how old I am?" and I wished I'd splurged and gotten bigger trees. After all these years of gardening, I am used to buying little and being patient and watching plants grow, and I don't very often think about how old we are. The distance I allowed between the little trees seemed vast, too, contributing to their skimpy display. But someday, probably after we're gone, their spacing will look just fine.

We have a good fruit nursery nearby, Windy Hill in Great Barrington, Massachusetts, and I managed to find there some old favorites and new appealing sorts. I brought home 'Mutzu' (also known as 'Crispin'), 'Gravenstein', 'Cox's Orange Pippin', and threw in 'Winecrisp' (a hybrid of 'Winesap' and 'Honeycrisp') as well as 'Northern Spy' and 'Macoun', the last two sorts heralded by friends, and praised by Rowan Jacobsen, the author of my favorite book on apples, *Apples of Uncommon Character*. Of the 'Macoun' (pronounced like cow), he advises that you "eat it under the tree" and swoon. "You can't take it with you," he explains, for " 'Macoun' neither ships nor stores—it functions like a seasonal madeleine in the memory of many a New Englander." Of 'Northern Spy', he says, "If I had to live on a desert island with only one apple variety, it might well be Northern Spy." Praise, indeed.

The new orchard

To the five apple trees, we added a pear called 'Magness', for the fruit of the existing old pears is barely edible, at least to us—as they drop to the ground, deer suddenly appear to hoover them up. We also added two peach trees, hearing that our neighbors had success growing them, imagining the bliss of pulling a ripe peach off your own tree. I know it will be a trick to get the fruit before the squirrels do. The final tree we acquired is a quince, a favorite of Bosco's for jam and paste. The fragrance of the quince fruit brought in to the warmth of a kitchen has never appealed to me, but I am recently won over by the deliciousness of the fruit when roasted for dinner or added to an apple crisp.

The nine purchased fruit trees were carefully planted in the sloping field the first spring after we moved in. A summer and fall of severe drought ensued. Much hauling of hoses, watering, and watering again while worrying that our well would go dry. Thankfully, it didn't. Most of the trees survived, only to be inundated in the past two years by torrents of rain, the year 2018 recording almost twice our average annual rainfall. Two of the little apples, 'Mutzu' and 'Winecrisp', were done in by nasty cases of cedar apple rust, exacerbated, I'm sure, by all the wet. The rust is an inevitable threat here since our woodland is wreathed at its edges with native red cedars, *Juniperus virginiana*. Because we are making every effort to be an organic garden, we need to rely on fruit trees that can survive pretty much on their own. I have replaced the 'Mutzu' and 'Winecrisp' with 'Macouns', for the one we planted so far seems particularly healthy despite the cedars.

Since we have no deer fence around our 17 acres, and the small trees with tender leaves and eventual fruit would be guaranteed to be caviar for deer, we had to protect them somehow. At first we just pounded in wooden stakes and wrapped wire around the trees, and after a while the stakes leaned this way and that, and the wire bulged, all rather hideous, and, of course, you could barely make out the trees through the mess.

On a recent trip to England while hurtling down a country road, I glimpsed a planting of young trees in a pasture, each protected with a generous square of fencing fashioned out of logs, 4 upright as posts, and 4 more logs attached horizontally as rails. I suspect they were constructed to keep the trees safe from grazing cows. I thought the enclosures handsome and decided they might be perfect for our orchard. Back home, I asked our beloved once-sometimes-twice-a-week helper John to gather 6-foot lengths of red cedar in our woods (there's no end to the half-dead red cedars there, shaded out for many years by lusty white pines). We decided on 4-foot-high posts, and made each square 6 by 6 feet with cedar railings nailed on top. The cedar barricades would keep the deer at bay, but I worried about fawns dipping in under the top railings, so we strung three almost invisible wire strands around the wood squares. Suddenly, our apple trees didn't look quite as insignificant. The new enclosures gave them a presence.

The autumn after the orchard was planted we started planting daffodils. This is the royal "we," for all I did was order them and scatter them about where I thought they should go, and then John planted them in the rough grass around the fruit trees. That fall, we put in 700 *Narcissus* 'Barrett Browning', a small cup chosen because it is a great naturalizer and has a delicate demeanor. At least that was what I thought we planted. The bulbs turned out to be a large-cup daffodil called 'Ceylon', a little less wild-looking than I would have wished. Seven hundred daffodils is a nice start, but on a good-sized hillside, several thousand is better. We're going about it slowly. Each year, John laboriously digs in several hundred more—the small-cup 'Barrett Browning', this time, correctly labeled, creamy 'Cheerfulness', with small double blooms, the fragrant tazetta 'Geranium', with clusters of small flowers, a simple white perianth around deep-orange cups. More important, we've added sweeps of the elegant pheasant's eye, or poet's narcissus, *N. poeticus* var. *recurvus*, which flowers after all the other daffodils have faded and thus extends the daffodil season. With its swept-back, satin-white petals and tiny white crimped cup lined in red with a green center, the poet's narcissus seems most at home in a wild setting.

Apple trees protected from the deer

Black-eyed Susans in the west field

Bosco makes sure the daffodils we force into bloom from the cold frame in the winter get added to the mix usually sometime in early July after their foliage has died back. *Narcissus* 'Hawera' and *N.* 'Minnow', both favorites for forcing, look particularly sweet growing in the wild. I'd forgotten that we even planted pots of forced spring leucojum in the high grass, and suddenly there they were this spring, flowering jauntily at the base of one of the apple trees.

I'm going to be a bit more extravagant in my ordering of daffodils for the orchard this autumn, having learned of a planting method a thousand times easier than digging with a shovel. My colleague Deb Munson plants her bulbs with a bulb augur attached to her drill, which makes the process not only fast but effortless. She uses it, too, when planting plugs in a meadow. John will be euphoric. He planted 200 plugs of meadow natives that I blithely ordered through the mail this spring. It was no easy task searching for empty spots among the tough grass roots to dig, sometimes with a shovel, sometimes with a pickaxe. Had we only known about the bulb augur.

When all the daffodils have finished blooming in the orchard, the meadow grasses are high enough to conceal their decaying leaves. And by the summer solstice, black-eyed Susans, ox-eyed daisies, and fleabane are flowering, peppering the soft grasses with yellow and white.

THE WILD
LAND

THE FIELDS

We had a small meadow at Duck Hill, not one that already existed but one painstakingly created according to the dictates of landscape designer Larry Weaner, famed for creating native meadows in these parts. After Bosco and I were married—this was 2000—we decided to build a pool in what had been a small horse and pony paddock when my kids were growing up. The empty horse barn was turned into an apartment and half of the paddock was ripped up to install a new septic system that could have serviced a high school. The other half was dug and regraded for the pool.

This left a weedy mess of a landscape. I didn't want yet another garden around the pool, which was long and narrow and served as our water feature as well as a swimming hole. And I certainly didn't want lawn, though that might have been a nice quiet bit of negative space in what was a busy garden. I decided instead, with romance in my head, to make a flowery field in this ½-acre patch surrounding the pool. How fine, I thought, to be swimming half hidden by tall grasses and a succession of wildflowers, forbs, as they are spoken of in meadow talk. I called in Larry to orchestrate this transformation, and approved his plan with some misgivings. I was vaguely uneasy that some of the flowers he suggested were native to Midwest prairies, not New England or New York, and others were cultivars of local natives rather than the straight species, but I finally thought, what the hell, let him do what he thinks best. Which meant at first spraying the entire half-acre with Roundup, a practice that fills me with horror today. Next came the planting of plugs and the sowing of seeds, and, in the following year we began to enjoy some native flowers among the weeds that immediately crept in. I had made another garden to weed.

We never really mastered those weeds, and the native grasses I craved didn't materialize. We did have tall grasses, albeit non-native,

The front field at Church House in spring

and some delightful flowerings here and there of penstemons, rud-beckias, helianthus, and asters. Canada goldenrod came in from the field beyond our property with a vengeance, and, though I love all goldenrods, this aggressive species was bound to take over the small acreage. By the time we sold Duck Hill, Japanese stilt grass, from who knows where, was stealthily infiltrating the small field, creeping among the stems of the grasses and forbs. A nightmare.

Finding a place with an existing meadow was high on my wish list when we were house hunting. Oh, the happiness, finally, of discovering Church House! I fell in love with the front field here—that soft, wide expanse of grassland dividing the lawn in front of the house from the deep east woods—the day in October when we first drove up the drive-way. It was tinged tawny pink, the thigh-high meadow grasses waving their seedheads gently in the breeze, and the place seemed serenely alive, a wild habitat, I soon learned, where birds love to swoop, feed, and nest.

That field and the west meadow that now laps up to the back of the house animate my days. Whether I'm walking our property or working in the garden, or indoors looking out a window, the movement, the life those fields offer, from wind and from the bird and insect life they attract, is my fond distraction. A few days ago—it is June—as I was talking to a friend by the cutting garden, I stopped mid-sentence, suddenly distracted by the sight of a small chipping sparrow landing on the tip of a slender blade of meadow grass in the orchard. The grass bent over dramatically with its weight, but, astonishingly, didn't break, and the little bird lingered there swaying, in search of a seed or tiny bug, I imagined, or just enjoying a swing.

Except for the orchard area, we do not mow the fields until spring, aiming to cut them soon after the snow has melted in April if the ground is not too wet. The tall grasses and dried forbs offer beautiful winter silhouettes, particularly etched against a snowy landscape, and provide seed for our winter birds. Because the orchard is full of daffodils that begin to shoot up in early spring, the meadow around the fruit trees is cut in early November, with any luck before it snows.

The high front field is comprised primarily of meadow grasses, and I have not meddled with it, loving its simplicity. Fescue is here from the days when it was a lawn, and several native grasses—little bluestem, tall redtop (*Tridens flavus*), big bluestem (*Andropogon gerardii*), and smoky purple love grass (*Eragrostis spectabilis*). I savor and encourage the little bluestem (with a mouthful of a Latin name, *Schizachyrium scoparium*), several large stretches of which paint this meadow pink and amber from late summer into winter. This fall, I'm going to try weed whacking patches of matted fescue right down to the ground and then scattering seed of little bluestem with the hope it may have the air and space to take hold.

Last July, much of the field was unexpectedly transformed by a sea of flowering Queen Anne's lace, and I found its effect too astonishingly pretty to fret about the fact that this wild carrot is not a native. I hadn't noticed much of any Queen Anne's lace in the field the year before. I wondered, what combination of weather and natural seeding had caused this enchanting show? Early goldenrod (*Solidago juncea*) and gray goldenrod (*S. nemoralis*) flower among the grasses in late summer, and I notice each year that more bergamot is appearing in the field. Meadows are not static any more than gardens are, and I will doubtless see different forbs weaving into the field's tapestry in years to come.

A 5-foot-wide path, mowed weekly, curves through the middle of the meadow, starting by a large sugar maple, then bending past a canoe birch, and straightening alongside a great boulder, which is covered in lichens and our tiny red native columbines that flower in May, prompting me to call it the columbine rock. The path then dips toward the woods, past a section of wet meadow not seen from the house, where taller goldenrods, wingstem (*Verbesina alternifolia*), New England asters, and Joe Pye weed flower in late summer, and finally curving into a grove of old white pines, its shadowy ground a bed of

Little bluestem

The columbine rock

fragrant needles. From here, you can turn right and descend into the deep, rich, east woods, or bear left and walk up the hill beneath oaks and an old cherry tree, back to the openness of the high grass field.

I am entranced by mown paths through meadows. I love the lure of them, wanting to know where they go and where they will end. I delight to the sense, once on the paths, of being dwarfed by tall grasses on each side, lost in their being. The contrast of textures such a path affords is a visual pleasure, the cut-velvet green band coursing through a feathery field.

In back of the house, what we call the west field is threaded by three such paths, in each case leading to a trail in the woods. The first of those threads in the meadow runs from a small arc of lawn by the cutting garden and living room terrace, straight north through the fruit trees in the orchard to that high viewing point where we have benches and the fire pit. You can rest here or continue on a path into the woods that separates us from our neighbors and is used mostly by them or us when visiting each other. The second of the three mowed paths is on direct axis with the French doors leading out from our mudroom, and from there beckons you on an adventure. It wends through black-eyed Susans and bergamot to the beginning of a trail leading up into what we call our northwest woods. The third and final path through the field begins on axis with the center gate of the cutting garden, and curves out of sight, providing a bit of mystery for anyone visiting the garden. If you follow the path, it eventually leads past our small quarry and ends at a trail into high rocky bluffs in the woods that are our most surprising habitat. A linking trail is mowed all the way around the edge of the meadow, and Bosco often uses that circular course to get some exercise, striding along with two walking sticks. I am hopeless at exercising that way, unable to keep from stopping along the way to cut down some interloper (the invasive bittersweet or knapweed) or inspect a flower or butterfly.

Unlike the front field, the back west field is naturally rich in flowers. Soon after the daffodils fade, the composites, or daisy sorts, start to bloom, most particularly our native golden black-eyed Susans. White ox-eye daisies are here, too, and the taller white fleabane, *Erigeron annuus*. By the end of June, the lovely blue spiked lobelia, *Lobelia spicata*, intermingles with the yellow and white daisies along with discs of white yarrow. Where the meadow laps the edge of the woods, thimbleweed opens its fragile-seeming white cups among the grasses. This charming native anemone, *Anemone virginiana*, was new to me when we moved here, and I love its common name, referring to its seedheads or fruit, which indeed look like tiny thimbles, and by autumn erupt into what seem like puffs of smoke. At the edges of the meadow, yellow sweet clover is flowering (*Melilotus officinalis*) along with its white cousin, and although both hark from Eurasia, they are not invasive here—that is, they do not seem to be aggressive enough to overwhelm or push out our meadow natives. Their tall, delicate spikes are delightful to cut for vases of roses and the first of the hydrangeas. *Lysimachia ciliata*, our native fringed yellow loosestrife, blooms in a shadowy, damp spot beneath one of the old pear trees, its stalks of nodding flowers clear yellow with the petals deepening to red at the base.

At the beginning of July, the dramatic stalks of milkweed (*Asclepias syriaca*) are opening their umbels of pale lavender-pink, and when passing by them you catch whiffs of their perfume in the air, beckoning you to stop and bury your nose in the scented blooms. By mid-July, bergamot (*Monarda fistulosa*) starts to flower, one of my favorite local natives, turning stretches of our back field into a haze of mauve-lavender with its soft shaggy heads. The flowers are actually clusters of tubular petals up close, beloved by hummingbirds and butterflies. The leaves of wild bergamot, like bee balm, are pleasingly scented, and sometimes in spring when I am walking in the meadow, my footsteps will stir up their delicious minty smell.

Path to the pine grove

By August, the goldenrods begin to flower, those graceful natives beloved by butterflies and birds that are often unfairly maligned for causing hay fever. Unfortunately, ragweed, which flowers inconspicuously at the same time, is the culprit. The rampant Canada goldenrod is not in much evidence in the back field except near the woodland edge, but the delicate early goldenrod (*Solidago juncea*) is here along with the charming flat-topped sort (*Euthamia graminifolia*), which I love to cut as a filler in summer bouquets.

August is when the little bluestem and Indian grass (*Sorghastrum nutans*) come into their own, and we have stretches of each in this field. The tall arching stems of Indian grass drip with feathery flowers, copper and yellow, turning buff in winter, and are striking seen against the evergreen background of white pines here.

I have introduced some natives that weren't here to the back field close to the cutting garden, reasoning that this was an area that just came into being because we stopped mowing. The showy goldenrod, *Solidago speciosa*, which I've admired in neighbors' meadows, is planted here now and I look forward to its handsome pyramidal plumes of flowers colored a soft yellow. Plugs of the smooth blue aster, *Aster laevis* (now annoyingly changed to *Symphyotrichum laeve*), were added to the field this spring. This aster is said to be particularly drought-tolerant and has loose panicles of pale blue-violet flowers. Butterfly weed, *Asclepias tuberosa*, which loves the dry, poor, quick-draining soil here, was woven in among the grasses a year ago, and it is now dazzling us with its deep orange jewel-like flowers. Like milkweed, it is a butterfly magnet, both as a host for their larvae and as nectar. I've tucked in some mountain mint, *Pycnanthemum muticum*, where the field is slightly moist, hoping it will spread. At Duck Hill I planted mountain mint in one of the garden beds, loving the effect of its silvery flower bracts that remain fresh and attractive all summer and fall. But if it's happy, it's hard to contain. Let it run in the meadow! This early spring, I sprinkled seed of the delicate

The trail at the edge of the west meadow

Indian grass in the west field

white *Penstemon digitalis* wherever there was a thin spot among the grasses, and maybe if we're lucky it will appear and flower next year.

As delightful as the flowers are, the high feathery grasses are what give the fields their distinct character. A few nights ago, before going to bed, I walked with our dog Sadie out into the back meadow. A thick fog had settled on the field, and all was pale and silvery beneath the night sky, the tall grasses just visible, etched softly through the blanketing mist. I suddenly realized fireflies were sparking the foggy night with their flashes of light, dancing above the grasses. Lightning bugs, we called them as children. As though this silent tableau was not captivating enough, all the while the air vibrated with the music of spring peepers, sometimes described as sounding like hundreds of sleigh bells. I lingered for a long time watching the misty field with its tiny fireworks, serenaded by the chorus of frogs.

Native columbines seeding in the rock bluffs

THE HIGH WOODED BLUFFS

The first winter we spent in Falls Village was especially snowy, for weeks at a time boasting an accumulation of a foot or more, requiring snowshoes to walk where plows or a shovel hadn't cleared the ground. We were not living in our house yet, but came every day and, if the sun was shining, I often lingered to explore the land. I wasn't at all clear at first where our boundaries were: The snow made it impractical to ask a surveyor to come mark the perimeters of our land, and I couldn't easily decipher the markings on the printed surveys we had. One particularly snowy day, early on, I ventured off on snowshoes through the back field and down into the woods beneath old white pines, passing what struck me as wonderful outcroppings of ledge rock. Continuing on, I came to a frozen pond and was astonished to find that I was standing below bluffs that rose almost three stories high above the frozen water. Icicles hung down the rock faces, which were half hidden by pines, cedars, oaks, and paper birch trees. Oh! I thought, is this possibly ours? Do we own this startling scene? Indeed we did, we do, and it is probably our most unique piece of woodland, dramatically littered with the limestone rocks and richly diverse plant life that are part of the remarkable Housatonic River valley landscape.

That pond is really a bowl-shaped depression that fills with water from winter snowfall and spring rains, a vernal pool that quickly dries up as warm weather settles in. The bluffs that are the backdrop to this pool are thought, possibly, to be the remains of another old limestone quarry, for some of the rock faces look cut rather than worn down by weather. In shadowy places, these faces are now softly clothed with mosses and bristle-leaf sedge, *Carex eburnea*, which favors such

Rock outcropping in our high woods

a calcareous (chalky) habitat. The slender, almost hairlike blades of this sedge spill and froth out of cracks and crevices, interrupted by the occasional tall, arching Solomon's seal that has managed to seed there, and the spoon-shaped leaves of ragwort, *Packera obovata*.

The dark gray rock ledges that so appealed to me that snowy day in spring are crusty with lichens and whitened with marble dust. They are home to a number of native wildflowers, with a succession of blooms from spring through fall. The tiny, dancing, red columbine, *Aquilegia canadensis*, is a favorite, flowering with abandon on and about the rocks in May. In summer, the equally diminutive harebell, *Campanula rotundifolia*, begins to bloom, pale violet-blue bells rising on wiry 8-inch stems seemingly right out of the rocks. It will continue to flower into fall.

One large flat ledge about halfway down the bluffs I call the picnic rock. It is smoothed from years of water spilling down over its surface, and sunbaked, positioned in a clearing in the woods with an open sky above and a surround of old red cedars and young oaks. I think of it as a place to have a summer basket lunch with some of our grandchildren—those who are old enough not to venture too close to its edge, for there is a precipitous drop into the vernal pool area some 20 feet below. Because of the sunlight here, tiny bushes of *Potentilla fruticosa* grow in the crevices, and the slender pale blue *Lobelia spicata* blooms freely in June along the edges of the rock. Blue-eyed grass, *Sisyrinchium angustifolium*, appears here earlier in spring, piercing stems above a fan of leaves like a miniature iris. But its flowers, barely the size of a dime, are simple and starlike, opening with the sun to reveal a striking purple that deepens toward a tiny yellow throat.

To one side of the picnic rock, in the shade of pines and oaks, the hop hornbeam, or ironwood, *Ostrya virginiana*, has colonized. Several have managed to grow tall, while others are still in their infancy or, in some cases, kept small by grazing deer. Hop hornbeam, I've discovered, is a small upland understory tree with an elegant habit and delicate birchlike leaves. Its appearance belies its toughness, for it survives

Carex eburnea and *Packera obovata*

Lobelia spicata and *Rudbeckia hirta*

in dry, stony shade on steep ridges. According to William Cullina (in *Native Trees, Shrubs, and Vines*), the wood is indeed as strong as iron, and was once used for airplane propellers and sleigh runners.

By the time the snow melted that first year, I was busy marking and clearing paths in the woods. At the direction of our new local doctor, I knew to wear high sox with my jeans tucked in, the shoes, sox, and legs of jeans then sprayed with permethrin to protect against ticks. With bright-colored surveyor's tape and Joyce Chen scissors in my pockets, I slowly wended my way through thickets of undergrowth, tying a length of tape to a shrub or tree where I thought a path should go. Sometimes I followed the narrow trails worn by deer, for they were often the best way to traverse a stretch of rocky land. If I was happy with the imagined path, I returned with loppers and secateurs to begin clearing enough to easily walk. Then John followed up, using a chainsaw where needed.

The middle path in the back field now continues into the northwest woods, rising gently through old white pines, aging junipers, or eastern red cedars as they are confusingly known, oaks, and the occasional sugar maple. The ground here is covered with the round-leaf ragwort (*Packera obovata*, formerly *Senecio obovatus*), also called groundsel and squaw weed, a native with a predilection for limestone woodlands. It runs by stolons, making glossy rosettes of serrated leaves the size of teaspoons, and it is surprisingly evergreen even in the harshest of winter weather. In early June, the woodland is lit with its yellow daisylike flowers that rise in clusters on stems 8 inches above the basal leaves. Our friend and great butterfly enthusiast Andy Brand came for a visit two years ago, and was thrilled to see all this ragwort, for it is an important butterfly host. Not because of its flowers, but for those oval-shaped leaves, which are the only food eaten by the larvae of the endangered northern metalmark butterfly. When the butterfly hatches, its preferred source of nectar is black-eyed Susans, of which we have a meadowful. I have yet to spot this small brown and orange butterfly but hope he has discovered us.

The trail in the northwest woods passes by another water feature—this time a curious small, natural pool that is almost always full of water, barring summer drought. A large American elm stands by one end (another thrill to discover), and at the other end a rock ledge juts out into the water wrapped with the roots of an old dying juniper. Just below the pool, a mysterious narrow pipe thrusts out of the earth—to drain the pool, we wonder? I found a battered tin bucket nearby. The pipe is surely clogged now; no water drips out of it. Paths are worn by animals' footprints on two sides of the pool, and I realize this is a favorite watering hole. I have decided to buy one of those outdoor motion-activated cameras to install here, for wouldn't it be wonderful to see what creatures come for a drink at night when we're asleep?

Beyond the pool, light streams in through the trees to the west, suggesting an opening, and ahead is the ravine. The trail now turns and begins to rise steeply, finally winding along the top of the bluffs, and you realize quite suddenly just how high up you are. Young chinkapin oaks (*Quercus muehlenbergii*) join white pines and red cedars along the way, their small trunks roughly barked and their long shiny leaves scalloped and toothed. The chinkapin oak is similar to the chestnut oak and is our native denizen of limestone ledges and rocky wooded hillsides. Dozens of seedling oaks cover the ground here above ragwort and Robin's plantain, *Erigeron pulchellus*, an asterlike perennial that flowers in early spring, pale violet daisies above paddle-shaped fuzzy leaves. In summer, the blue-stemmed goldenrod, *Solidago caesia*, flowers here, for it is a woodlander, its short yellow spikes clustered in the axils of its leafy stems. Just about the only fern you see in our high wooded bluffs is the ebony spleenwort, *Asplenium platyneuron*. It is a charming thing, quite small and slender, its once-divided pinnate leaf standing straight up like a tiny soldier. This delicate fern only occurs in calcareous woodland, happily existing on rocky ledges and mossy banks.

Having reached the top of the bluffs, I used to skid down the steep hillside to the picnic rock, and again, below to the vernal pool. Bosco

Ebony spleenwort

Cedar steps and pearly everlasting in the high woods

rarely ventured here. But now, John has fashioned rustic steps out of cedar logs set into the slopes with a longer log serving as a banister to guide our way. The steps are already graced with ragwort, and the creeping stems of dollar-leaf or prostrate tick trefoil, *Desmodium rotundifolium*, which, indeed, has 3-part leaves as round as silver dollars, and dainty racemes of mauve-pink flowers in August and September.

THE LOW EAST WOODS

Surrounded as our house is by a flat plain of fields, it isn't at first evident that Church House stands halfway up a considerable hill—not until you climb up to the bluffs in the west woods, or, conversely, follow the path through the east field down into the pine grove and descend from there into the east woods. The trail falls steeply past moss-covered boulders into a world of ferns and Pennsylvania sedge, of skunk cabbage in the spring, of witch hazels, red maples, elms, and hemlocks. What a contrast these two habitats are, one rock-littered, high and dry, the other low, rich, and moist. Rocks are here in this low wood too, ledges and outcroppings seeming to hold the banks as the land drops away, and, below, where the ground becomes level, stone walls that tell the tale of an earlier time.

My real introduction to our east woods was on a February walk with the forester and surveyor Matt Kiefer, whom I asked to come mark our boundaries. He told me that ours was a young wood, and explained that the white pine, eastern red cedar, birch, and white ash that are prevalent here are "pioneer trees," quickly taking over what he suspects was farmland in the nineteenth and early twentieth centuries. Those telltale stone walls, running both north to south and east to west, are simply made up of huge lichened rocks piled against each other, suggesting that much of this woodland was once pasture. I learned from Tom Wessels's marvelously logical book, *Reading the Forested Landscape*, that smaller stones mixed in with large ones in a wall suggest the boundaries of cultivated land, for plowing in New England inevitably turns up a lot of stone that has to be cleared. But here, no smaller stones were added to the walls. The discovery of strands of barbed wire attached to half-rotted cedar posts confirmed that animals were pastured here, cows or perhaps sheep.

Skunk cabbage in the wet woods in May

Part of a pasture wall

Some of our biggest trees in the east woods grow along these walls—sugar maple, white ash, black cherry, and white pine—100 or more years old. The great limbs of some of the ash trees are grayed, smooth, and bare with death, and I look around at the other still-healthy ashes scattered throughout this wood and wonder what it will look like when they're gone, for the devastating emerald borer is approaching our area, and these great trees are already succumbing to fungal diseases.

Old red cedars are also remnants of an earlier time when the land was open pasture, for they thrive in open sun. I think of them as our own Italian cypress when I see their narrow, pyramidal stance along highways and in fields. Here, they are shaded out by the fast-growing white pines, except at the wood's perimeter where, facing the sun, they are still richly clothed with blue-green needles and blue fruit (actually cones) that the birds love. Where the ground is thickly wooded, many stand tall but dead, just snags now, their trunks free of their shaggy bark, smooth, fluted, muscular.

Some of our great pines have immense trunks rising about 10 feet, then forking into several heavy leaders, suggesting that they, too, were once in an open setting. For, the white pines that seed and grow in the open, as in a pasture, are often attacked by a weevil that kills their leader and causes this forking. We have a gathering of pines at the bottom of our east woods, however, that grow tall and straight like the ship masts they were once prized for. This spot is near the edge of an open wetland, brightened by shafts of sunlight. The ground beneath the trees is blanketed with their needles, and the air is heady with their fragrance in the warmth of summer, transporting me back to childhood summers in New Hampshire, walking barefoot on pine needle paths, or sitting beneath pines with my aunt in Massachusetts. Soft grassy clumps of Pennsylvania sedge are established here, and evergreen partridgeberry, *Mitchella repens*, threads through the needles, often decorated with its scarlet fruit. I call this place the pine cathedral.

Matt introduced me to prickly ash, *Zanthoxylum americanum*, also known as the toothache tree, which dominates the shrub layer in our east woods. It seems to grow in groves, and is quite tall and willowy, its branches well decorated with thorns. Matt said it only grows in limey soils and is native here, the northernmost New World species of the citrus family. Oil extracts from its bark were once apparently used medicinally (for toothaches among other ailments, I assume). Best of all, prickly ash is the primary larval food source for the giant swallowtail butterfly. I caught my first sight of this huge jewel-like butterfly this summer, coming to rest on a flowering head of Joe Pye weed in our wetland. By late August, the giant swallowtail is everywhere in our garden, often in pairs or threes, fluttering among the flowers searching for nectar. How satisfying to know we have enough prickly ash for a whole lot of swallowtail caterpillar meals.

At our farthest east boundary, the ground can be quite wet, especially in a winter and spring like this past one, when we had an inordinate amount of rain. A trail I made there is not always passable, but when it is, or when I have boots on, I love to walk there. A grove of our native witch hazel, *Hamamelis virginiana*, fills the understory along the path and is beautiful at all seasons, with the grace of its spreading limbs and its handsome, fresh green, ovate leaves. In the fall, those leaves turn a soft yellow that appears as sunlight in the woodland understory, followed by the astonishing flowering just before winter of spidery clusters of yellow threads along its branches. A few hemlocks are here and white canoe birches among the witch hazels, and one large, shiny black birch, which was new to me. If you take a knife and peel a small piece of bark from the trunk of the black birch, it smells of wintergreen, and, indeed, oil of wintergreen is distilled from the bark to flavor chewing gum and candy.

Our path here curves around several tall American elms, the mossy base of their trunks splayed out to anchor them in the wet soil. I have to look up so high to see their leafy canopy, I often lose my balance. Ramps

Giant swallowtail butterflies

Rue anemone and Jack in the pulpit

(wild onions) grow here and trilliums, our native ruby-red *Trillium erectum*. In May, too, there are starflowers, *Trientalis borealis*, and carpets of rue anemones, *Thalictrum thalictroides*. Jack in the pulpits, *Arisaema triphyllum*, appear suddenly through ferns and by fallen logs. Miterwort flowers appear then, their tiny white cups fringed like snowflakes on a two-leafed stem. Pale blue violets appear on hummocks, and, later, wild geraniums open their mauve-pink blooms.

Two springs ago, I was walking along this path with my son Keith and his wife, Ali, admiring wildflowers, when suddenly we came upon a large black bear. Keith intelligently surmised it was a mother bear, for she didn't turn and lumber off as most black bears will when startled but stood her ground, in essence confronting us. Suspecting there were cubs nearby, we backed up quietly and quickly, returning the way we came. But my foolish and fearless son went back into the woods later that day with his camera (he is an avid birder, and carries an appropriate lens for capturing distant fleeting subjects). He found the mother bear standing in the same place by a large white pine, so he looked up and spotted two cubs among its high limbs. The mother bear then methodically climbed the tree, pulling her massive body up along the trunk with her strong, clawed front paws. The great pine, it seems, was a babysitting tree, somewhere for the cubs to be safe while she searched for food on the ground. I didn't walk down the path again until autumn, for fear of accidently getting between that mom and her cubs.

Sometimes I am envious of the mature deciduous woods that some of our friends have, with beech trees and great old oaks, an understory of mountain laurel and moosewood, with soil acid enough for an outer fringe of blueberries. But then I think how alive our young limestone woods are, how diverse they are, full of history and character, how rich in plants and experiences that are new to me. And I see them as a wonderful adventure, and am content.

THE FEN

Our deep low wood opens onto a small wetland along its southern edge. Not deeply wet, but with an inch of water running between hummocks of shrubs and wildflowers—except in a dry summer, you want boots on to walk through it. In spring, the wooded ground along the edges of the wetland oozes with mud your boots sink into as you weave through a carpet of skunk cabbage to reach it. Sometimes my sock-covered foot comes out of my boot, the boot staying lodged in the mud. When you arrive at the small wetland, you are surrounded by an enthralling place, a circle wide open to the sky and sunlight in contrast to the shadowy woodland from which you have just emerged. Almost no trees are here in the opening, just occasional thickets of willows and shrub dogwoods.

Late last fall and again this summer, I had the good fortune of walking in our east woods and wetland with Robert Reimer, a professional gardener and new friend who has an encyclopedic knowledge of native plants. It was he who first pointed out the elms (I had not noticed them!), and showed me how their leafy branches differ from birch or our native hornbeam because of their zigzag arrangement. I learned how to recognize a red maple merely by the cross-hatching of the fissures on its trunk. As we walked along the paths, I asked Robert the names of various ferns we were passing, for I only seemed to know the common Christmas fern, the sensitive fern, and the little polypody. He taught me how to differentiate the majestic and ancient interrupted fern (*Osmunda claytoniana*) from its close cousin, the cinnamon fern (*O. cinnamomea*) by looking at their backsides to see if there are tufts of small rusty hairs at the base of each leaflet (yes, with the cinnamon fern, no with the interrupted). He pointed out the difference between the lady ferns (*Anthyrium*) and the wood ferns (*Dryopteris*), how the wood fern is sharper looking, not quite as graceful or delicate. And if you break open their leaf stalks (or stipa), you notice that the lady fern has only two interior fibers, called bundles, whereas the stem of the wood fern has several. He introduced me to sedges and grasses, like the dangler, *Carex gracillima*, and the towering wood reed grass, *Cinna arundinacea*, in the damp woodland. And when I wondered how you tell the difference between a grass and a sedge, he memorably said, "Sedges have edges," for, at their base, their triangular stems have three distinct sides, whereas grasses have rounded, hollow stems.

When we reached the wetland, I asked him what I should call it. It's not a bog, is it? I asked. "Heavens no," he said, "it's a fen." A fen! Oh, how romantic that sounded, straight out of a nineteenth-century English novel, much finer than a bog or a swamp. A rich fen, at that, he added. What exactly *is* a fen, I wanted to know. Robert explained that it is an open, calcareous wetland, with a seepage of water, rich in minerals and therefore rich in plant species. He thinks we (he) could probably find 100 native species in this small area.

On both occasions, autumn and summer, we tromped through our fen as he pointed out one wildling after another. I, all the while, wishing I had a pencil and paper, trying desperately instead to record accurately on my phone. That first late autumn day, we found evidence of red swamp currant (*Ribes hirtellum*), *Geum rivale*, shrubby cinquefoil (*Potentilla fruticosa*), *Iris versicolor*, marsh fern, and cinnamon fern. In July, the fen was in flower with pink swamp milkweed (*Asclepias incarnata*), white boneset (*Eupatorium perfoliatum*), tall, rose-purple Joe Pye weed (*Eupatorium maculata*), blue vervain (*Verbena hastata*), meadowsweet (*Spirea latifolia*), and another mountain mint (*Pycnanthemum virginianum*). Robert showed me the porcupine sedge and a native mint, pennywort, and water horehound. In a particularly wet spot in the fen, great hummocks of tussock grass (*Carex stricta*), sitting high on their self-made islands, made for an astonishing sight, and Robert told me bog turtles like to nest in them, so you have to be careful when you're walking there.

The red osier dogwood, *Cornus sericea*, is the dominant shrub in the fen, its cranberry-colored stems often kept low by browsing deer. The silky dogwood (*C. amomum*) is here too, heavy with clusters of fruit that turn blue in fall. Three sorts of willows lend some height to the fen, among them *Salix discolor*, the American pussy willow we all know and love, with leaves that are dark green above and felty white on the underside. *Salix sericea*, the silky willow, is smaller and finer in appearance, and the diamond willow, *Salix bebbiana*, is distinguished by its coarser, crinkly leaves and an abundance of catkins that develop later than those of our common pussy willow.

By late September, the panicled aster, *Aster lanceolatus*, is flowering, white and pale violet, all through the fen. Swamp maples and American hornbeam lean into the fen from our surrounding woods, as well as winterberry, and, in their fiery autumn dress, add to the spectacle. The hornbeam, *Carpinus caroliniana*, is another native plant new to me, called musclewood because of its smooth sinewy trunk. A close relative of the hop hornbeam, it prefers the rich moist ground found in our low woods.

My dream is to have a boardwalk that traverses the fen. Then I will walk it every day, seeing what is in flower, watching for birdlife. A wide worn trail, probably an old farm road, runs along one of our stone walls at the foot of the hill dropping into the east woods. If you follow along the road, you come to the edge of the fen by a venerable eastern cottonwood, *Populus deltoides*, so tall we see the upper half of it from the house. I will start the boardwalk there, curving it past red osier dogwoods, out into the open, then across the wetland to the far wood's edge, continuing it around through the skunk cabbage–littered wet woods to higher, drier ground and my path. Garden friend Robin Zitter, who has experience with boardwalks, suggests simply using pairs of 8-foot locust planks from a local saw mill, each pair nailed onto three big logs placed beneath them to elevate the walkway. I think I need to do it soon. A winter's project perhaps.

The fen in October

WORKING
WITH NATURE

A NEW WAY OF
THINKING

Among the plants we managed to bring from Duck Hill was a division of a double white trillium that Bosco and I were given as a wedding present by a good gardening friend. Without giving it much thought, I decided to plant this precious gift down in our low east woods, reasoning that trilliums grow here naturally—at least our wake robin, or purple trillium, *Trillium erectum*, does—and therefore it would be happy there. I was certainly right about the suitability of the moist rich woodland soil. I just didn't think how it might be at home in another sense. The glorious white trillium, *T. grandiflorum*, doesn't grow in our woods, and, in fact, is not native in Connecticut, though it has apparently naturalized in a few areas. Its double form, which looks in flower like a hothouse gardenia, was bound to stick out in our woods, an overdressed stranger at the wrong party. Had I thought it through, I would have planted the double trillium in an intimate, humus-enriched spot in the front garden in the shade of one of our shad trees, or beneath the fothergillas among hellebores, where this gardenesque wildling might look more at home.

But I carried our prize down into the east woods and planted it at the side of the path by a moss-covered rock where ferns and purple trilliums abound. I thrust a white label in the dirt next to it, and promptly forgot about it. The next spring, I noticed the label in passing with no sign of a white double trillium. I assumed it didn't make it through our winter.

Fast forward to the following spring. The trillium appeared but didn't bloom. But, on a subsequent walk when spring was warming into summer, I came to a screeching halt beside the plant and its label. In horror, I saw Japanese stilt grass (*Microstegium vimineum*) growing in a circle all around it. Oh my god, I thought, I am going to be responsible

Church House woodland

for bringing Japanese stilt grass to northwest Connecticut! I rushed back to the house for a plastic bag, returned and fell on my knees, pulling out the delicate bamboo-like stalks of this sinister invader.

Now, each time I pass this spot, I stare at the ground for a while, looking for signs of the stilt grass. I usually find a few. In New York and southern Connecticut, I've seen Japanese stilt grass spread through open woodland like a plague, until it covers the entire floor. Even if I find no signs of it here today, I know to continue being vigilant. Although the grass is an annual, its seed can stay viable many years in the soil. This nightmare episode forced me to think about our woodland in a different light. The knowledge that I introduced a plant unknown to these woods, in a pot of foreign soil contaminated with the seed of a virulent exotic, made me shudder.

I've never owned wild land before, I mean land now in a natural state, uncultivated. We had a woodland walk at Duck Hill, but it was a fantasy as such, merely a copse of trees along the south and west borders of our 3-acre property where I created paths and planted shade-loving plants—epimediums, hellebores, primroses, spring ephemerals, native woodland phlox, anemones, violets, early bulbs. And yes, trilliums, all of which flourished and spread with abandon in this small shadowy garden. Our wedding gift of a double white trillium tripled in size. It didn't look out of place.

I loved that woodland path and probably miss it most of all the various garden areas at Duck Hill. When we first came to Church House, at about the same moment I naively planted that white trillium, the idea crossed my mind to take a small section on either side of the pathway that leads to our neighbors in the upper woodland and make it into a garden like the one I missed at Duck Hill. After the trillium–stilt grass incident, I thought better of it. For our small stretches of forest have a character of their own, and it became clear to me that I wanted to listen to the land, discover its denizens, thrill to what it offers, nurture it—not make it into something else.

The woodland path at Duck Hill

FIGHTING INVASIVES

When I think how often the understory of our northeastern woodland is overtaken by Japanese barberry—the prickly bushes untouched by deer and able to spread unhampered to the detriment of our native plants—I am over the moon to find that we have almost none here. I have no idea why we should be this lucky. Occasionally I'll come upon a surprise bush, and, if it's small, I easily pull it up by hand. A larger branched bit is dispensed with using a spade. Blessedly, we have almost no winged euonymus in our woodland either. I came across one good-sized specimen deep in the east woods, which we quickly dug out, but otherwise I merely notice a jagged twigged seedling once in a while and give it a tug. November is the perfect month to eye the woodland floor for any lurking sprigs of each, for their leaves linger and turn vivid colors—rust-red in the case of barberry, bright pink with the euonymus. This is the time of year when we are most aware of their aggressive spread throughout the northeast, along with Norway maples, overwhelming and thugging out our native woodland flora.

Though we have virtually no barberry or euonymus here, we are not free of unwanted invaders. I have declared war on three major invasive exotics in our woods—bush honeysuckle (*Lonicera morrowii*), European buckthorn (*Rhamnus cathartica*), and, last and by far the worst, Oriental bittersweet (*Celastrus orbiculatus*). The honeysuckle is seemingly everywhere, at the edges of the woods as well as in its shadowy interior, forming dense thickets at times where the woods are open, and, in the process, certainly crowding out native shrubs. It seems to grow with abandon whether the woods are high and dry or low and damp. It is an arching 5- to 6-foot shrub of unremarkable countenance, with small, typical cream-white flowers followed by red fruit that the birds eat and then disperse by seed. The honeysuckle pulls up easily if it's not too big, and we do that if we are not dislodging native wildflowers in the process. Much of the time, John and I cut it to the ground. Of course it sometimes resprouts and we cut it again. I actually prize the cut branches from older bushes, especially branches that are forked, for the wood is quite strong and they make great garden stakes.

Common, or European, buckthorn, the second invasive we work to expel, is new to me. I gather it was introduced to these parts in the 1890s as an ornamental shrub for hedges (as was the multiflora rose), and, of course, it escaped. The black fruit is consumed by birds and the seed scattered, and there you are. It is apparently able to grow in a variety of soils, but has a penchant for woodland soil with a high pH. To make matters worse, its roots and leaves are thought to have an allelopathic effect on the surrounding ground, chemically suppressing other plant growth. Older specimens are easy to recognize in the woods because of their scaly, rough, dark bark and shiny, dark green, oval leaves. Cutting them down usually requires a chain saw, for they are often the size of a small tree in height, and John is appointed to deal with them. I can easily shear to the ground any leafy sprouts that appear later on.

A lot of the work clearing these two invasives—bush honeysuckle and European buckthorn—happens in winter, when John doesn't have a lot of garden work to do. We soon had to come to grips with what to do with the huge amount of branches and brush we accumulate. The woods are too steep to think of hauling the brush away, or bringing in a chipping machine. John and I decided to designate certain spots in the woods for brush piles, thinking these would be good habitats (probably for the rabbits that are eating my garden, among other creatures) and would decompose naturally over the years. What I didn't know is that John would make them beautiful. He lays the branches in teepee fashion, and as the piles build up, they begin to look like houses a beaver would meticulously craft. Land art, I think of them.

Clearing our woods is a big job, probably 10 acres in extent, and John and I work on it together once a week, weather willing. I designate a certain area, along a certain path, and together we point out what is to be cut. When John first started to work for us four years ago, he knew nothing about native and non-native plants. But he has a quick and discerning eye, and now knows the trees and shrubs, good and bad, that make up our woods. Even in winter, he will recognize a gray dogwood before I do, just by the grace of its twigs.

Occasionally, we will decide to cut down a dead eastern red cedar or pine that is in our way or leaning in a threatening fashion. But mostly we leave the dead trees standing as part of the habitat, offering food and nests for birds and other small furry creatures. John saves the straight cedar trunks that he downs and piles them to one side for future fence posts, rails, or steps. I am always astonished by the color of the heart of a cedar, a vibrant sienna red, which must explain its common name. Where dead trees have fallen that don't affect our walking paths, we leave them to rot, and enjoy seeing them covered with mosses and ferns.

Bittersweet is my nightmare. This third invasive is everywhere in the woodland and along its edges, even creeping into our fields. We battle it mostly by cutting it to the ground, or pulling it up if we can. I know we will never be rid of it, but we are making an attempt to keep it at bay. An alternative is to cut and brush the stems with glysophate (Roundup). But that's hundreds of stems and consequently a lot of poison, and I shake my head. Our first concern, of course, is to make sure no vines are climbing into trees or native shrubs. When we first started working in the woods, some of the bittersweet trunks were 2 and 3 inches in diameter, wrapping boa-constrictor fashion around young trees, their bark ominously snakelike in its pattern.

There is no time of year when we're not pulling and cutting the vines. It is July as I write this, and suddenly new, fresh green tendrils of bittersweet are rising up from the ground everywhere, looking for, and often finding, something to wrap around. The first warmth of summer seems to give them new energy, and it is now that we have to be most vigilant. Even if we've cut the bastards down in early spring, we have to go back at it now. When I walk along our paths in the woods, I invariably take a pair of loppers with me, and cut and cut as I proceed. In a few very severe infestations where nothing much else is growing, John cuts them to the ground with a strong weed whacker. One theory is that aggressive natives will take hold and weaken the bittersweet eventually. I hope this is true.

We do have a swath of the invasive *Rosa multiflora*, famed for colonizing unmowed fields, where our stretch of wet meadow meets the tree line and the fen. But once a year, this area is bush-hogged, which keeps the rose more or less at bay. This year, I noticed an ominous patch of phragmites that's crept into this meadow from our neighbor's property, and I feel challenged how best to get rid of it. We will cut it down, but I know that won't do much good. Not sure digging it out works either. I might have to resort to brushing the cuts with Roundup. We just can't let it get into the fen, for that would be this wetland's ruin. The fen is now, astonishingly, quite pure; it is almost entirely free of non-natives.

We have three troublemakers in the high fields. Spotted knapweed, *Centaurea stoebe*, from Europe, has invaded small sections of the front and back fields, loving our increased heat and dryness in deep summer. I even find bits of it growing right out of ledge rock in the bluffs. Every year it seems to be increasing, and now, as much as possible, we yank it out. It is quite attractive in bloom, creating a haze of lavender with its bachelor-buttonlike flowers, but it is thugging out precious natives that enjoy the same poor, dry soil—little bluestem and blue-eyed grass (*Sisyrinchium angustifolium*), for instance, and *Campanula rotundifolia*.

Mugwort (*Artemisia vulgaris*) is a nasty invasive that we see along the roadsides in summer. It wants to wreak havoc in the east field. So far we have only two patches of it and we've been weed whacking it down to the ground once or twice a summer. But now I'm going to try spraying it with an organic weed suppressor that won't kill the grasses around it. At least that might weaken the mugwort's resolve.

I've seen whorled bedstraw, *Galium mollugo*, which is native to Eurasia, take over whole fields. I am surprised that it's not yet on the

Connecticut Invasive Plant List put out by the University of Connecticut. Its tiny white flowers in densely branched clusters have a dainty charm and I often cut sprigs of it to mix with stouter flowers in a vase. But there's nothing dainty about its habit. We only have small swaths of this bedstraw growing among the meadow grasses at this point, and I make some attempt to pull it out when it is in flower so it won't go to seed. I think I have to become a little more serious about battling it.

If all this sounds like an almost hopeless struggle, it is. The operative word is almost. Even the slightest bit of effort ridding the land of invasives means additional light, air, and ground for native plants, and that means more birds, bees, and butterflies. Encouraging news comes from Penn State University where a team spent seven years repeatedly removing invasive shrubs from a 40-acre wood. Native plant regeneration and diversity far exceeded their expectations.

Will I have that much time? Will the next owners of this land care? Who knows.

One of John's wood piles

An oak seedling in our high woods

NURTURING OAKS AND WITCH HAZELS

Doug Tallamy writes in his seminal book, *Bringing Nature Home,* that an oak is the most powerful plant you can add to your landscape in terms of wildlife species it supports. Well, we don't have to add any oaks here, just protect and nurture the ones we have. Our friend Robert Reimer says our high rocky woodland wants to be an oak savannah, the ground is so thickly populated with oak saplings. They are mostly chinkapin oaks here, maturing to rough, scaly barked, slender trees with narrow, toothed leaves; the occasional burr oak is here too, with more deeply lobed leaves. In the low east woods, we have swamp white oak, *Quercus bicolor,* the back sides of its leaves felty white, as well as chinkapins, and one or two tall red oaks.

I was spoiled at Duck Hill, for the entire garden was deer-fenced and, consequently, I never had to think of protecting plants from their browsing. We have no deer fence here, though sometimes I dream of one. I doubt I could afford to fence in 17 acres, and I would hate to have barriers between us and our neighbors, let alone the nuisance of a gate across the driveway. But how wonderful it would be to watch seedlings of native trees and shrubs grow up in the woodland unchecked by the deers' destructive appetites.

Small seedlings of our native witch hazel, *Hamamelis virginiana,* can be found coming up in parts of our damp low woods. The sight of their wavy ovate leaves thrills me, and I long to see them grow up. The gray dogwood, *Cornus racemosa,* is here too, 3 or 4 inches high, easily recognized in leaf, as all dogwoods are by their distinctively handsome veining, the leaves often tinged with red. At our wood's edge, where it meets the fields, we have some high bushes of this dogwood that have somehow escaped the deers' notice. Now I would like to see whole colonies of it in the newly opened areas of our low woods where we've cleared a lot of honeysuckle. This native shrub, sometimes called the red-panicled dogwood, is attractive in all seasons thanks in part to its fine twigginess and billowy habit, and also because it flowers on pedicels that become a deep rose red and support white fruits in late summer. The white berries are relished by birds and disappear quickly, but the colored pedicels remain, decorating the branch tips into winter.

I know these seedlings will never get any bigger if I don't protect them, for in winter the deer munch any young twiggy growth they find. We do not have many deer in the woods in the summertime, maybe an occasional buck or a doe with a fawn. But as soon as hunting season starts and the nature preserves around us open up to hunters, the deer flood our property along with our neighbor's 70 acres that surround us on two sides. And, having found a haven, they stick around all winter.

Once a month, I walk through our woods and spray the oaks, witch hazels, and dogwoods with an organic, smelly deterrent called Deer Defeat. It is helping. But I also plan to put up some netting or wire around areas where the witch hazels are or where I think an oak has the space and the purchase of open sky to grow up. It won't be pretty, but if I can get trees and shrubs high enough not to be browsed, the effort will be worth it.

We have a young grove of native chokecherry here, *Prunus virginiana*, that I also spray against deer browse. The birds love the fruit of this shrubby understory tree, and even more important, its foliage, like that of the black cherry, supports the larvae of hundreds of butterflies, moths, and skippers. As Doug Tallamy wisely reminds us, if you want butterflies, you have to make butterflies. We need to plant and protect native species of trees, shrubs, and wildflowers because they serve as host plants for butterfly larvae and because the birds that we love are dependent on those caterpillars as a source of protein to feed their young. I was told recently not to rid our crabapples of tent caterpillars— as I have blithely done for years, wiping them out of the crotch of the tree with newspaper and dumping the whole ugly mess in a garbage bag. Turns out, those caterpillars do little harm, and are food for a number of bird species as well as bats. Tallamy writes that our garden-variety crabapple trees are so similar in leaf chemistry to the rare native species that our native insects can't seem to tell the difference and therefore are happy to use them as hosts.

By the fen, small plants of our native spice bush, *Lindera benzoin*, are struggling to survive the deer. Seedling American hornbeams are here also. They just need some protection until they grow up, and then the deer will leave them alone. The same holds true for the occasional pagoda dogwood, *Cornus alternifolia*, which I am overjoyed to find in the woods, and the young hop hornbeams by the bluffs. These two elegant trees deserve netting too. Oh, for that deer fence.

Black-eyed Susans and fleabane meet the wood's edge

NATIVE SPECIES ADDED TO THE FIELD'S EDGE

Looking out our kitchen windows, I felt early on a desire for more autumn color where the back field washes up against the woods. Goldenrod is there in summer, and the charming lavender, heart-leaved aster in September, but there are few understory shrubs, and the trees you see from this vantage point are mostly evergreen with trunks bare of foliage—high-pruned white pines and old, shaggy barked red junipers. Since I'd already fiddled with this back meadow, adding to its display of native wildflowers, I thought I would continue by planting a few appropriate shrubs and small trees at the edge of the field to transition through that middle ground and give us some late season interest. Because I was intending to choose native species, I knew they would add to the diversity of what we can offer birds and butterflies.

Right off the bat, I planted a sassafras tree. I've never had one before, but am enchanted by its curious mitten leaves. Actually three sorts of leaves can exist on a tree, entire (unlobed), 3-lobed, and the 1-thumb mitten. Its layered habit, rather like a dogwood tree, will be perfect in front of the woods, and its fall display is said to be fiery with orange, yellow, and scarlet. The leaves are a primary food source for the larvae of the spicebush swallowtail butterfly, and the blue-black fruit of the female sassafras is favored by migrating birds. What's not to like? A sassafras that is a respectable size is apparently difficult to transplant, but of course I planted a small sapling, barely 6 feet high, which has a good chance of flourishing. We just might have a few years to wait before it dazzles us. The tree is known to sucker, and I have visions of a sassafras colony at the edge of our field.

Our friend Robin Zitter gave us two small native hornbeams (*Carpinus caroliniana*) from her nearby woods. We might have added them to the edge of the fen where I've discovered they exist naturally, but instead we dug them in beneath arching pines at this woodland-field edge where the land dips down and I hope will provide them with some of the moisture they love. Like the sassafras, the musclewood's habit is wider than high, with outstretched limbs and leaves that are also brilliantly colorful in the fall. Their elegantly toothed oval leaves provide larval food for the tiger swallowtail and white admiral butterflies.

Next spring, I hope to add a hazelnut to the mix, *Corylus americana*, which is a large, spreading understory shrub also celebrated for its fall color. It intrigues me even more for its winter catkins, its early spring flowers (the male flowers long and yellow, the female, rosy nubs), and particularly its distinctive nuts. I've only seen pictures of them growing, but each nut is wrapped in a pair of fringed and ruffled green bracts, and a small cluster of them looks like a sumptuously ornate blossom. Those nuts when ripe are gobbled up by squirrels and other small mammals, as well as turkeys and grouse.

We've planted several bushes of witherod (*Viburnum nudum* var. *cassinoides*) at the edge of the field that I hope are getting enough moisture. Robert and I spotted a witherod growing in the fen this summer, and, seeing it in its natural setting made me question my placement at the edge of dry upland woods. But it is said to do all right in a moderately dry soil and, so far, despite astonishing summer heat, the bushes seem healthy. This shrub is a native prized for its edible fruit, which can vary in a cluster from green to pink to blue to black, often several colors present at once. Witherod's lustrous leaves turn deep red to fire-engine red in the fall, when migrating birds take advantage of its fruit.

Hobblebush (*Viburnum lantanoides*) is a viburnum I didn't know before coming here. If you drive up Canaan Mountain in the early spring, you'll see colonies of it in flower among ferns along the edges of the woods. The wheels of large white flowerlike sepals around central green-white discs are startling in their showiness, and the heavily ribbed, paddle-shaped leaves are beautiful in themselves. I reasoned that if they flourished at the wood's edge ten minutes away, they were bound to thrive here. Wrong. Several small specimens were bought and planted, and, after two winters, only one of them is still alive, barely. I didn't take into account that the woods on Canaan Mountain are acid-loving and we are on lime. Nor did I know that they transplant with difficulty. I will baby the one that has survived and hope for the best. If you have a moist shaded wood, moderately acid in makeup, this is a must-have gem of a viburnum.

A few bushes of black chokeberry, *Aronia melanocarpa*, now grace our field's edge. We had a small colony of this chokeberry at the end of the woodland walk at Duck Hill and I am pleased to have it again. This native is rather delicate in habit to about 5 feet with clusters of small white flowers full of bees and then purple-black fruit that lingers into winter, offering late foraging for birds. The shiny, elliptical leaves turn a rich red in fall. This plant is another naturally existing in wetlands but adapts to a dryer situation. There's a native red chokeberry, *A. arbutifolia*, that is slightly showier with red fruit and brilliant fall foliage, and I am tempted to add a few bushes of it to the perimeter of the field and wood. Nurseries usually offer an improved cultivar with larger fruit called 'Brilliantissima'.

The question often comes up whether cultivars of native plants are as effective for promoting wildlife as the straight species. According to Doug Tallamy, some cultivars are, but not all. When the cultivar's leaf color is different from that of the species, red- or purple-leaved, for instance, the native is spurned as a host plant. Apparently the chemical makeup of the plant changes when you fiddle with the leaf color. If, in a cultivar, the size of the fruit is enlarged for a flashier effect, the berries can be difficult for some birds to swallow. When single-flowered shrubs and perennials are made double, that can also be a problem. Consider *Hydrangea arborescens* 'Annabelle', for example. Unlike the straight species where fertile flowers are surrounded by decorative sepals, the fat round blooms of 'Annabelle' are entirely made up of sterile bracts, which is no good for bees.

I happened to notice in July, when our swamp rose was in bloom, what a magnet it was for bumblebees. Every 5-petaled flower no bigger than a silver dollar shuddered slightly with the machinations of one fat bee after another, buzzing loudly as each wallowed in the flower's golden stamens. Our well-loved, repeat-blooming double-flowered roses, such as 'Therese Bugnet' and 'Stanwell Perpetual', attract Japanese beetles but few bees. In a public radio interview with garden writer and blogger Margaret Roach, Tallamy said he wished more nurseries would give us the choice of buying straight species of native plants, for they offer the maximum value ecologically.

I've often preferred the species of certain perennials over "improved" cultivars, but for aesthetic reasons. Inevitably, a species is more graceful, or taller (I do love tall flowers), or more fragrant. Until recently, I didn't think much about the importance of what a species offers in the way of food for wildlife. Reading Doug Tallamy changed that. Bugs have taken on a whole new light. I no longer just think of plants as decoration, but consider them as food too. Now, when I see the leaves of a native flower or shrub with holes in it or its edges chewed to bits, I think, oh, well, maybe that caterpillar was nourishment for some freshly hatched baby birds, and, if it escaped that destiny, then perhaps it will turn into a heavenly butterfly.

TAKING HOLD

Starflower in the low woods

A DIFFERENT EMPHASIS

The gardens we've made in the past four years, the echoes of Duck Hill we've introduced, give us much pleasure: cold frames, greenhouse, orchard, gravel paths and terraces, the flower borders to walk through, and perennials and flowering shrubs to grow and cut. All of this enriches our days. But daily I find I am drawn away from the gardens to the wild land. I cannot easily explain why the fields and woods we've inherited call to me in a way the flower gardens do not. Perhaps it is precisely because they're not cultivated, they hold mysteries, they offer discoveries unknown to me. I thrill to the unexpectedness of flowerings almost every day in spring, summer, and fall as I walk through the meadows and along the woodland paths, many that send me to books to find their names. I'm beguiled by the traces and sightings of denizens that share our land, not only birds, but bears and bobcats, coyotes and turkeys. Less happily, of course, is the evidence of deer, the damage they wreak, although I couldn't wait to show Bosco a twelve-point rack of an antler I discovered lying on our forest floor last autumn, the horn satin-smooth from rubbing against tree trunks, striated creamy white and tan. It now has a place of honor on a shelf, as much a treasure to us as a beautifully crafted bird's nest or their jewel-like eggs discovered abandoned in the grass.

Before coming to Church House, I never thought much about habitats and ecosystems and biodiversity—that is, as far as our own acreage was concerned. Like Falls Village, our former town of North Salem is exemplary in having large tracts of unspoiled land in preservation, and I felt fortunate to be able to open the back gate of our garden at Duck Hill and walk out into those fields and woods. I wrote small pamphlets about some of the wildflowers that flourished there. For many years, I listened, enthralled, to the song of a meadowlark in the large meadow that abutted our property and knew how fragile the existence was of this

Dahlias along the walk to Duck Hill's chicken house.

ground-nesting bird due to haying and development. I tried as best I could to have a thoroughly organic garden there, and made sure to plant a healthy percentage of native species. As with most country gardens, we had a variety of microclimates, and we were lucky to have areas of both shade and sun. But, in contrast to our new home, every inch of Duck Hill's 3 acres was cultivated and contrived.

Here, at Church House, we've stumbled on a landscape rich in ecological value thanks to its diversity of natural habitats. Deep, rich woods feeding into wetland; low, damp meadow rising to fields high and dry; upland, rock-littered forest. Yes, our 17 acres have been monkeyed with over the centuries, first possibly by Indians burning and clearing, then by settlers, and finally by farmers. But only traces of their inhabitance are apparent now, and the young woods show inspiring signs of regeneration. This is as close as I'll be in my lifetime to stewarding a piece of land in its wild state. And it has changed me as a gardener.

I have a new hunger to learn. About wildflowers, about calcareous habitats, about ecosystems, about birds and pollinators. It is humbling and at the same time exciting to know that education never ends when your passion is working out of doors. We are always learning. After almost sixty years of digging and planting in a garden, I still make mistakes and am wiser from them. With the best of nurseries and knowledgeable gardeners around me, I am still introduced to and beguiled by new intriguing plants. But now, for the first time, I am making acquaintance with plants that already exist in their natural settings on land we own, however briefly: Indian pipe and bishop's cap in the woods, curious oaks, musclewood, and hop hornbeam overhead, tussock sedge in the fen, spiked lobelia and thimbleweed in the meadow. My new challenge and delight is educating myself about what is here in this wildness, as well as protecting and savoring it.

The beauty of moss in our woods

BIRD LIFE (NO CATS)

One September morning just before dawn, nine months after we moved into Church House, I heard the soft hooting of an owl outside our bedroom window. A series of four sonorous hoots, the second and third more rapid than the others, repeated for about half an hour as I lay in bed, almost holding my breath with the delight of it, until all was abruptly quiet as the sky became light. At breakfast, I listened to Audubon recordings of owls on my iPad to learn I'd been serenaded by a great horned owl. I believe he is the largest of our local owls, with a fine pair of upright tufted ears and a reputation for consuming an impressive daily quantity of small furred creatures such as voles and rabbits. Go to it, I say, as a litany of rodent destruction—devoured perennials, annuals, bulbs, even shrubs—runs through my head.

Our great horned owl continues to wake me just often enough to be a welcome excitement, perched in our ancient sugar maple a few feet from our bed, during all the cold months from late September through February. I've shaken Bosco awake once or twice to hear his hooting, but mostly I don't disturb my husband because he is a sound sleeper and naturally a night person rather than a morning one. Once or twice I've heard our owl hooting in the evening, answered by a mate deep in our woods. My hope is that they have appropriated one of the huge red-tailed hawk's nests high up in the trees here. For a Christmas recently, my son Keith built me a wooden nesting box made to order for great horned owls, and, at great risk to life and limb, managed, with the help of his brother-in-law, to haul it up to a high crotch of a white pine at the edge of our back field. He then dressed it appropriately with twigs and branches. No sightings have occurred at it, and we suspect this lovingly conjured nest is still not high enough to appeal to our owl. We've also learned they're not keen to nest in a white pine. I long to catch a glimpse of him, but haven't dared venture outdoors when he is hooting, for fear of scaring him away. And, anyway, it is always dark when he lets his presence be known.

Our barred owls are not so shy and call to each other in the morning and afternoon throughout the year: *who-cooks-for-you-all* they'll hoot from our low east woods as I go about my work in the garden, and it always makes me smile. I've only spotted one once, not here, but fifteen minutes away, at Bartholomew's Cobble in Ashley Falls, Massachusetts, a nature preserve famed for its quartz-marble bedrock and outcrops that are host to an extraordinary collection of rare calcium-loving plants. The owl was perched on a low tree branch there above the Housatonic River, silently staring at us, its dark eyes circled with what looked like black-rimmed goggles, surrounded by a fat ruff of brown- and white-striped feathers.

Other birds call from the woods as I work outside that I rarely, if ever, see. The flutelike song of the wood thrush is the most thrilling to hear drifting up from our east woods on quiet mornings and evenings in the spring and early summer. Until we came to Church House, I had not heard a wood thrush since I was a young girl spending summers at our family lake house in New Hampshire. The damp pine and beech forest behind our house there came alive each morning with his beautiful trilling. Not at all melodious but easily recognized is the song of the ovenbird, a repeated *teacher-teacher-teacher*, which I hear all spring long from the interior of our woods. I would love to come across its nest on the forest floor, a domed oven made of dry leaves and grass with a side entrance.

We often hear the squawking of a pileated woodpecker as it flies from tree to tree. I stopped to watch one doing just that on a March afternoon this spring while walking down into the east woods. It finally settled on an old black cherry tree in front of me, and became silent, its scarlet crest vividly displayed as it hopped up and down and around the trunk, pecking for ants. The impressive rectangular holes this oversized woodpecker makes decorate quite a few dead cedars in our low woods, sometimes with a circle of fresh shavings on the ground beneath. Once, when I was weeding in the cutting garden, a pileated woodpecker swooped out into the field and began pecking in the grass, which surprised me, and then flew to an old stump nearby, all the while making a racket with its repeated, high-pitched call. I thought this performance rather curious for a bird with a reputation of being shy.

We have two vessels for water in the cutting garden: one, the shallow birdbath in its center, and the second, a granite bowl placed on the stone wall just above the east beds. Watching the birds come here from the surrounding field is a summer delight—robins, goldfinches, bluebirds, catbirds, sparrows. Sometimes they just perch on the edge of the water and delicately take several long drinks before flying off. But often, after quenching their thirst, they settle into the water and begin wildly splashing and dipping. This goes on for many minutes, the robins bathing with the most abandon, until, refreshed, they stand up and shake the beads of water from their feathers. One or two birds are often waiting on a flower stalk or on the fence for the next turn.

On a recent August day, my son Keith and I sat in the pool garden and, sharing binoculars, watched an eastern wood-pewee in action. It would choose a perch—a high bare twig of a Japanese lilac at the edge of the meadow for a while—and make short flying spurts into the air, catch an insect, and return to the exact same perch. This was repeated over and over again. A small gray bird with a pale throat and chest and a jaunty crested cap, the wood-pewee has an easily recognized call, a clear whistle of *pee-a-wee*, that rises higher in pitch at the end. I often hear this little flycatcher when I don't see it.

Because we live in an area inhabited by black bears, we put away all bird feeders in March when it seems likely hibernation is over. Bears have the keenest sense of smell of all our local fauna, and whiffs of birdseed and suet will bring them to our doors. The usual spring birds arrive even without the lure of feeders. Suddenly one March day, a flock of robins will appear, peppering our lawns, hopping, pausing, cocking their heads sideways, looking for worms. Along with cardinals, their song seems the essence of springtime. Carolina wrens, chickadees, chipping sparrows, and song sparrows are my constant companions in the early spring garden. Soon, bluebirds and tree swallows come to nest in the birdhouses in our east field. We begin to hear the sweet phoebe calling his name.

One early May morning this year, I had my first sighting of the season of our ruby-throated hummingbird. A brilliantly iridescent fellow zoomed into the west terrace garden as I was having breakfast, and systematically went from one 'Hawera' daffodil to another, at least thirty of them that I had planted in a waving row, dipping into each small nodding cup until he had visited them all. He then zipped across the terrace to drink from the Virginia bluebells, clumps of which were in flower in the shade of a viburnum. This hummingbird is a daily visitor to our garden through spring and summer, and I always know when it is nearby because of the *burring* noise of its wings. I was alerted to one this morning in the cutting garden and watched while a little female hovered briefly before piercing the drooping pink catkins of kiss-me-over-the-garden-gate. She then shot over to the 'Lemon Queen' sunflower, looking for nectar in its tiny disk flowers, and moved on to a vivid red coneflower, attacking its disc as though it were a pin cushion. I have a fervent wish to find a hummingbird's nest one day, a tiny delicate cup made of woven grasses and spiderwebs lined with down.

Our first spring at Church House, a pair of Baltimore orioles arrived as the apple trees came into flower and stayed to make one of their extraordinary nests—a sack hanging precariously from the tip of a high branch near the crown of our great old black cherry tree. Last year, we had no orioles visit. Was it a coincidence that there were no apple blossoms either? This spring, our great old apple tree was covered in pink-tinged flowers, and passing underneath it I heard a different chirping from within the boughs. Keith, who is a passionate birder in his free time, had just texted me to be on the lookout for orioles. And here he was, smaller than a robin, black and orange, singing, feasting on the apple blossoms (or was it caterpillars?) above me. I heard his mate rustling and answering him in an adjacent crabapple. At Keith's suggestion, I placed cut oranges on a rock by the garden to entice them to stay. But I haven't heard them singing this summer and I peer up into the branches of our high maples, ash, and cherry, but see no sacks hanging down.

All my life I have had cats. Until now. Our last feline—the only one Bosco ever knew—was a living nightmare, a black and white fellow named Felix, more often referred to, once he was no longer a kitten, as psycho-cat or the cat-from-hell. He brought grandchildren to tears, sitting innocently, eyes blinking, even purring, as they approached to pat him, then suddenly swiping at them with claws extended. He did the same to me in winter when I was getting dressed to go out for an evening, pulling on a pair of stockings. Wham. Stocking ruined. A spot of blood dribbling down my leg. He deliberately pushed crockery off tables, reducing to smithereens a favorite Dresden porcelain teapot I inherited from my family. He raised his tail and peed on chair corners and sprayed a fanciful brass clock of Bosco's, turning it green. He never washed himself, and preferred sleeping on the finest white bed linen, leaving a shadow of grime behind him. I had towels draped all over the place.

Worst of all, the cat was a serial killer. We had a dog door so Roux and Noodle, our resident terrier and dachshund at the time, could come in and out from the garden. Felix learned to use that flapping door to bring his small furry creatures indoors. I would wake up to a rabbit hopping on the bedroom floor above ours, the cat-from-hell cruelly prolonging his death. He'd bring a mouse into the kitchen, still alive, sprawl out on the kitchen floor, drop the mouse, and lazily watch as it scurried under the stove or into the heating vents. One morning I came in to breakfast to find a rat, thankfully dead, left for my delight. He killed many birds, lying in wait by the feeder, then pouncing. That was as unforgiveable as scaring the grandchildren.

Birdbath in the cutting garden

I tried to find a barn that would take him. No luck there. I suggested to my daughter, Jean, a veterinarian, that she end his life. But her conscience rightly won't let her put down an animal unless it is hopelessly ill or in distress. Thank the lord, Felix died of old age just before we moved to Church House.

Bosco thought all cats were like this. They're not. I wish he could know a sweet, companionable puss, like many my family has had over the years. But the fact remains that the majority of cats, when allowed outdoors, will kill birds. They are natural hunters. There are reports that cats kill over two billion birds a year in the United States, though the accuracy of that number is, not unreasonably, in question. Our bird life here at Church House is rich and joyous, and I cannot bear the thought of introducing such a threat to their existence. Nor would we want a cat that only knows the indoors of a house.

Dogs, yes, we will always have dogs. Life without a dog is unendurable. But we will not have a cat here.

SHARING A HOME

Duck Hill was always mine. When Bosco and I got married in 2000, I had lived there for nineteen years. The gardens were established, the boxwoods shoulder high, the flowers in satisfying drifts. The house was my cozy nest. It was my haven, my solace, the place where I loved most to be.

It's not easy to move into someone else's home. Especially if at least one of you is possessive. One of Bosco's pleasures has always been working in a garden, but it was hard for him at first to find a niche in mine. I thought of the garden at Duck Hill as my work of art, however imperfect. I was forever editing, attempting to create an atmosphere, more interested in painting pictures with plants than acquiring them. Bosco lives to acquire plants, loves nothing more than accumulating and propagating plants, never bothering much about where they're going to go in the scheme of things. Design is not his thing. It's the doing that he loves.

We are both nesters, which is probably one nester too many. And, at this point in our lives, we are pretty opinionated about how things should be done. I load the dishwasher a certain way, Bosco changes my arrangement, thinking his more efficient. I confess to doing the same in the reverse. He likes a lot of knickknacks on our tables, I prefer them cleaner—although I don't hesitate to pile them with books. Outdoors, Bosco is all for more variety in the flower borders. I want drifts and puddles of a few plants. "Too many this's and that's," runs through my head, the observation of a dear colleague about a particularly busy garden.

The dust gradually settled during those first years of our marriage, and we learned to accommodate and appreciate each other's ways. During our time at Duck Hill, we developed new areas of our land—the woodland path, the vegetable garden, the pool enclosure,

The Boscotel at Duck Hill

the orchard. Together we built the Greek Revival chicken house, and Bosco enthusiastically took over caring for our brood—chickens, ducks, turkeys—saving every green or sweet scrap from the kitchen for their delectation, gathering eggs in return. We built the Boscotel so he would have a room of his own. He had his greenhouse. He had a pool.

But the garden was always thought of and known as mine. The old farmhouse, too. Bosco never got used to its low ceilings and walls of books everywhere, and, although his pictures were hung where there was room, and his furniture and bibelots woven in, he never loved the house the way I did. It wasn't hard for him to leave.

How fine then to end up in a place we discover together, neither his nor mine, but ours. I think we love it equally. Daily we pinch ourselves about living here, about how lucky we are. Each window, each door in the house frames a view of the hills and the sky, and we'll call to each other to come look, at the fleeting light, at the shadows, at the colors. In summer, when we linger on the terrace into dusk and the sunlight abandons the fields around us, suddenly Canaan Mountain is lit golden lavender, or flaming orange, and we say, Oh, oh, look, isn't it wonderful.

We have our routine in the garden. Bosco spends a lot of time potting and repotting. He fusses over his plants indoors and out, growing pineapple lilies (varieties of eucomis), acidanthera, tuberose, and gloriosa lilies in pots to stage on our terrace in summer. He tends his figs, with John's help, bringing them out from the garage in spring as they begin to leaf out, then watching for fruit, hoping our summer will last long enough to ripen them, sometimes pruning the trees in the fall before they go back indoors. Along with agapanthus and clerodendrons, the figs spend the winter in the barely heated garage where Bosco watches over them. His greenhouse and his collections of begonias and geraniums are his particular joy. He loves nothing more than discovering new sorts, propagating them, and giving rooted bits away to friends.

Meanwhile, in spring and summer I am off with my trug and my kneeling pad to weed. I am one of those odd creatures who actually enjoys

Sadie in the back field

Kitchen bouquet

weeding. I find it utterly absorbing, on my hands and knees stirring the earth, pulling out interlopers, looking at flowers and leaves up close, their patterns, their fragrance, familiarizing myself with their habit and what they like or don't like. And then, standing up, as I need to do often now, I have the instant gratification of seeing what I've accomplished.

I do a lot of standing and staring at the garden, looking at combinations of colors, textures, and patterns, deciding what works, what doesn't, trying to make pictures as I did at Duck Hill. I change my mind dozens of times. This is the beauty and the hazard of designing your own garden. My garden notebook is full of directives of what to plant more of, what to discard, what to move, sometimes crossed out later with a note saying, "No, leave it." Bosco rarely sees the garden's imperfections, and is ever appreciative, annoyed sometimes when I share my criticisms of it, merely wanting me to join him in enjoying its moments of beauty. Which, of course, I do.

Afternoons, I often go for a walk in the woods. Sometimes I end up there for hours, engrossed with what I see, what is there, lingering to battle whatever threatens the native treasures I've discovered. Bosco's knees are bothering him so he rarely goes with me, but I tell him of my discoveries when I return and show him pictures on my phone. We both like to walk on the paths in our fields, and our beloved Sadie follows us, making detours into the high grass to hunt for voles. She is our first rescue, and Bosco likes to say she was put together by a prestigious committee. She has the face of a terrier, the whiskers and front legs of a shih-tzu, the streaked tan and black coat of a German shepherd, and the tail of a fox. We think she is the best dog we've ever had.

We entertain a fair amount—Bosco is a people person and adores company, and I love to cook. Our old place was affectionately dubbed the Duck Hilton for a number of years. He teases me, complains really, because I'll invariably spend half the day arranging flowers before friends come for dinner, and then cook the rest of the day, with the result that I am often beat by the time our company arrives. But it is what I love to do. Bosco always notices when fresh gatherings of flowers are in the rooms and, bless his heart, is quick to praise them. We are both overjoyed when our children (two of his, four of mine) and grandchildren (fourteen in all), who are scattered all over the United States and Europe, come to stay. They seem to like it here too.

And so, we coexist in our new home, happily, gratefully. We know our years are limited and think how lucky we are to be here today in this shared place we cherish.

CHANGE

I've always preferred the old and familiar, whether a beloved home or a well-worn sweater or a dog-chewed pair of boots. And yet, I have on occasion chosen dramatic changes in my life. I packed an overnight bag and abruptly left college one spring day when my family was in crisis, and never returned to complete my education. I walked away from two marriages when I found them insupportable. I left a home where I said I wanted to live until I died.

Bosco experienced unthinkable changes in his life. As a young boy in Hungary, he lost his home and all belongings. He grew up in America mostly separate from his family. In middle age, he lost his wife to cancer.

Those traumatic twists and turns, those losses, have, against all odds, brought us to a life together, a treasured companionship, in a new home, a new place. And we know that change can be good.

A friend, who has a stunning garden nearby, primarily of shrubs, conifers, and flowering trees, asked me the name of my new book. When I told him, he said, "It is a topic of great complexity so many of us face—transplanting with judicious root pruning in the process." I think of Bosco, unpotting his old figs, carefully cutting back their roots, then replanting them in fresh soil in a new pot. Root pruning a plant is said to help with transplanting, by encouraging new feeder roots and temporarily slowing down top growth.

We shed attachments, moving here. In that uprooting, I've also slowly shed a way of thinking as a gardener—partly because of my age, partly because of the magic of this place. In the first year here, I was intent on acquiring shrubs and perennials that I loved at Duck Hill and missed. I now realize I don't need all of those echoes. Some of the exotics I planted stick out like sore thumbs here, and I am considering giving them away. We left behind three intensive acres of gardens. I feel no need for a showcase now. In the years I have left, I simply want to savor the outdoors and garden for the joy of it.

I know the flower gardens we now have are bound to become too much for me to take care of, and I'm already scheming for that eventuality. The pool garden is not a worry for it is mostly hydrangeas, and the new ones I've planted will knit together when full grown and only require a ground cover at most. Out will go the daylilies, the sanguisorbas, the dahlias. Hydrangeas only need pruning once a year, if that. As for the front garden, I can ease out needy perennials and replace them with low shrubs.

It's the cutting garden, curiously, that is truly high-maintenance, being comprised primarily of an ever-changing tapestry of annuals from spring through fall. I am fortunate now to have a young woman to help me weed a few hours once a week, and I will continue to enjoy that help as long as we can afford it. But the cutting garden can just be returned to meadow, if need be.

I remember seeing a picture of the designer Fernando Caruncho's garden in a magazine. The original plantings by his house had been felled by disease, as I recall, and, in replacement, he simply seeded the whole space with cosmos. Long-stemmed, gracefully swaying, pink and white, single-flowered cosmos. It looked enchanting. I could do that too, when the cutting garden becomes too much to handle.

When I can no longer garden for very long, I will still have the fields and the woods to walk in. My new roots, my heart, are spreading deep in that wild land, and the life it affords that we are helping support. Bosco asked me the other day, why we have plants in the front garden that also grow in the wild here. Wasn't it repetitive? he said. Why not instead have more plants that are unusual here. He was referring to the Culver's root, sunflowers, ironweed, and asters woven into the beds. I answered, Because they are beautiful up close, and because they bring the wildness to our doorstep, the bees, the giant swallowtails, the goldfinches.

Cosmos in the cutting garden

Helianthus 'Lemon Queen'

I am sitting in my third floor study looking out at the sugar maple almost in touching distance from my window, its grayed, flaking, furrowed bark streaked with black-green moss, its wide branches mottled with crisp patches of pale green lichen. The deeply cut maple leaves are stirring in the breeze, and I can just see our grassy field and red barn through them. If I turn my head, I catch glimpses of our hills, misty blue, gently rolling above the canopies of other trees. In the distance, I hear the hooting of a barred owl.

Like many of us, in the last few years, I am often consumed by the news, deeply troubled by the inequities, hatred, and tragedy tearing at our country, upset by the hypocrisy in our government, and the looming disaster of climate change. How lucky I am to have a place where I can sometimes lose myself among trees and flowers and birds, and forget the troubles of our world.

There is a round hole in one of the burls of the old sugar maple where a catbird hatched babies this spring just opposite my window. I watched as she (or was it he?) approached the hole, her beak tightly grasping what seemed like a bristly collection of insects, and little heads appeared, mouths open. While I was working in the front garden, the catbird often perched nearby and mewed at me, telling me that she was busy, don't disturb, she had work to do. Later that spring, I found a young catbird dead in the grass, and wondered if it had fallen out of the tree. Undeterred, the parents went on to have another brood, this time safely in the depths of the old boxwood bush beside our front steps. I hope those young ones will return to serenade me next spring.

ACKNOWLEDGMENTS

First, I am most grateful to Tom Fischer and Andrew Beckman for spurring me on to write this book, and to all those at Timber Press who carefully and patiently brought it to fruition. Warm thanks to Laurie Dunham, Inge Heckel, Bunny Williams, and John Rosselli, as well as my family for making our move to Church House memorable in the best of ways. To Edward Burlingame, who knew the name of my book before I did, and cheered me on during the writing process, as did Ruth Skovron, Margaret Roach, Elise Lufkin, Frances Palmer, Karen Pinson, Keith Dickey, Kim Dickey, and Laura Palmer—thank you all. I am indebted to Matt Kiefer, Andy Brand, Jeff Lynch, and Robert Reimer, who taught me about our native flora, and Tricia van Oers, Andy, and my son Keith, who identified the bird song in our fields and woods. Thank you Deb Munson, Robin Zitter, Katie Ritter, and Marilyn Young for bringing back to Church House some of the plants I loved at Duck Hill. And thanks to Bob Wright, Danielle Giulian, and John Gualan for the work you've done to make our land more beautiful. It is my deep pleasure to include Ngoc Minh Ngo and Marion Brenner, whose gorgeous images of Duck Hill and Church House grace these pages, and, finally, Bosco Schell, my partner in this new adventure, for his never-ending support.

PHOTOGRAPHY
CREDITS

All photographs are by Ngoc Minh Ngo except for the following:
Marion Brenner, pages 10, 13, 16, 56, 90, 93, 183, 204, and 216

INDEX

Rothermere, Claudia, 88
round-leaf ragwort, 160
Roundup (herbicide), 140, 186, 187
Roux (author's dog), 45, 51, 212
'Ruby Spice' summersweet, 122
ruby-throated hummingbird, 211
Rudbeckia hirta, 159
Rudbeckia subtomentosa 'Henry Eilers', 62
rue anemone, 171, 172
Russian sage, 58, 59

S

Sackville-West, Vita, 116
Sadie (author's dog), 78–79, 219, 221
Salix bebbiana, 175
Salix discolor, 175
Salix ×erythroflexuosa, 123
Salix gracilistyla 'Melanostachys', 123
Salix gracilistyla 'Mount Aso', 123
Salix sericea, 175
Salvia 'Argentina Skies', 85
Salvia guaranitica 'Black and Blue', 61
'Sangria' geum, 80–81
Sanguisorba canadensis, 70
Sanguisorba hakusanensis, 70
Sanguisorba obtusa, 70
Sanguisorba tenuifolia, 70
Sanguisorba tenuifolia 'Henk Gerritson', 80
sassafras, 196
Scabiosa ochroleuca, 98
scarlet curly willow, 123
scarlet runner bean, 89
scarlet tasselflower, 84
Schizachyrium scoparium, 143
scilla, 94

Scilla siberica 'Alba', 103
Scutellaria incana, 62
seasons. *See specific season names*
sedges and grasses, 174. *See also specific plant names*
'Segovia' daffodil, 65, 105
self-seeding plants, 88, 92–99, 143
Senecio obovatus, 160
serviceberries, 51
shadblows, 51, 61, 114
sheepberry, 126
Shirley poppies, 94, 95–97
showy goldenrod, 150
shrub roses, 68, 116
shrubs, 114–127
silky dogwood, 175
silky willow, 175
Sisyrinchium angustifolium, 157
skunk cabbage, 166, 173
smooth blue aster, 150
'Snipe' narcissus, 104
snow, 20, 155
snowdrops, 64
soil pH, 118, 122, 126, 185, 198
Solidago caesia, 161
Solidago juncea, 143, 150
Solidago nemoralis, 143
Solidago speciosa, 150
Solomon's seal, 157
Sorghastrum nutans, 150
spice bush, 193
spicebush swallowtail butterfly, 196
spotted knapweed, 187
spring season
    birds in, 210–211
    bulbs in, 64–65, 100–106
    at Church House, 20–23